Fatal Isolation

Fatal Isolation

The Devastating Paris Heat Wave of 2003

RICHARD C. KELLER

The University of Chicago Press
Chicago and London

Richard C. Keller is professor in the Department of Medical History and Bioethics at the University of Wisconsin–Madison. He is the author of *Colonial Madness: Psychiatry in French North Africa*, also published by the University of Chicago Press, and editor of *Unconscious Dominions: Psychoanalysis, Colonial Trauma, and Global Sovereignties.*

The University of Chicago Press, Chicago 60637
The University of Chicago Press, Ltd., London

Printed in the United States of America

24 23 22 21 20 19 18 17 16 15 1 2 3 4 5

ISBN-13: 978-0-226-25111-0 (cloth)
ISBN-13: 978-0-226-25643-6 (e-book)
DOI: 10.7208/chicago/9780226256436.001.0001

Library of Congress Cataloging-in-Publication Data

Keller, Richard C. (Richard Charles), 1969– author.
Fatal isolation : the devastating Paris heat wave of 2003 / Richard C. Keller.
pages cm
Includes bibliographical references and index.
ISBN 978-0-226-25111-0 (cloth : alkaline paper) — ISBN 0-226-25111-X (cloth : alkaline paper) — ISBN 978-0-226-25643-6 (e-book) — ISBN 0-226-25643-X (e-book) 1. Natural disasters—France—Paris—History—21st century. 2. Heat waves (Meteorology)—France—Paris—History—21st century. 3. Disaster victims—France—Paris. 4. Paris (France)—History—21st century. I. Title.
GB5011.48.K45 2015
363.34′92—dc23

2014036980

♾ This paper meets the requirements of ANSI/NISO Z39.48–1992 (Permanence of Paper).

For Max

Contents

Introduction

Situated roughly a dozen kilometers southeast of central Paris, the suburb of Thiais lies about midway between Orly Airport and Rungis Market. It appears at first like many of the depressed communities of the *banlieue*, the peri-urban region of the Paris suburbs. The four-lane D60 national highway separates it from its neighboring suburb of Chevilly; several buses along this corridor link the southeastern suburbs with the Paris Métro's terminus at Villejuif. The highway itself is lined with big-box stores, filling stations, and fast-food chains, punctuated by stands of couscous and Greek sandwich restaurants. But as the highway nears Thiais it becomes clear that another industry predominates in the town of 28,000. Florists, funeral services, and monument producers abound, bordering a vast open space that sits behind a forbidding gate on the southern side of the highway, the Parisian public cemetery of Thiais.

Unlike the better-known Montparnasse and Père-Lachaise cemeteries in Paris—the burial places of Jean-Paul Sartre, Simone de Beauvoir, Oscar Wilde, and Jim Morrison among dozens of other celebrities—a sizable portion of the cemetery of Thiais is dedicated to those buried at public expense. At the southern extreme of the cemetery lie four divisions that are colloquially known as the *secteur d'indigents* or "poor section." Each contains about 200 identical individual tombs, labeled with engraved brass tags bearing the victim's name, along with birth and death dates. In each division, several rows of tombs remain unlabeled, indicating that their occupants are unidentified. These plots are owned by the city of Paris and serve as temporary repositories for the unclaimed bodies of those who die within the city's limits or in several suburbs. Many of these graves are filled with the bodies of the homeless, buried at public expense when no family member claims them. Others buried here include individuals with few or no social contacts, who were found dead in their apartments and had no established burial plans, as well as those who died in care of charitable organizations such as the Petits Frères des Pauvres. Others still—and in far fewer numbers—are those who specifically requested burial there out of solidarity with the poor. Family members can claim bodies in these tombs for up to five years after burial; after that point, the authorities at the Paris morgue (the Institut Médico-Légal) may cremate the remains at any time in order to make space for new arrivals.[1]

At the time of this writing, roughly a hundred bodies in divisions 57 and 58 at Thiais are those of victims of the heat wave of 2003, the worst natural disaster in contemporary French history (see figs. 1 and 2).[2] A catastrophe linked to a high pressure front that remained static over western Europe for weeks in August of that year and assailed much of the Continent with record-high temperatures, it was a disaster that inflicted the greatest damage on the poorest and most isolated populations in France, as these bodies suggest. They are the so-called abandoned or forgotten victims of the disaster, whose bodies remained unclaimed by family. They are those who died (and to a great extent lived) unnoticed by their neighbors, only discovered in some cases weeks after their deaths. And they are those who rapidly became the symbols of the disaster for a nation wringing its hands over the mismanagement of the heat wave and the social and political dysfunctions it revealed.

The devastating heat wave or *canicule* that swept through western and

FIGURE 1. Division 58 in the Thiais cemetery. Photo by the author.

FIGURE 2. Tombs in Division 58, Thiais. Photo by the author.

central Europe in August 2003 was by every measure an extreme event. The disaster hit France particularly hard. Daytime highs reached 40°C (104°F) in Paris for days on end. More important, evening minimum temperatures dipped only to the low 20s C (low 70s F), giving bodies little respite from the heat even at night. Ozone pollution levels compounded the heat's effects, and the duration of these temperatures for two weeks without relief made the climate insufferable. Catenary wires supplying electricity to high-speed trains melted, stalling travelers for hours in the heat with no power. The drought that France had experienced since February devastated trees to the extent that many Paris parks—the limited green space that might have protected some from the heat's worst effects—had to be closed because of the danger of falling branches. The foundations of stone houses cracked because of the dehydration and settling of surrounding soil. The country suffered some four billion euros in agricultural losses. But far and away the greatest damage was the heat wave's human toll: roughly 70,000 lives lost to the heat in Europe, and some 15,000 in France alone.

For the news media, for public officials, and for French citizens aghast at the heat wave's merciless toll, however, one aspect of the disaster was particularly horrific. Weeks after the peak of mortality, hundreds of unclaimed bodies lay in France's makeshift morgues, which included refrigerated trucks parked in the Paris *banlieue* and the food warehouses at Rungis Market.[3] These "forgotten" bodies rapidly became the symbol of aging and isolation in the modern West, and a repository of national shame with respect to both the origins and the management of the catastrophe. The publication of the names of the dead and their collective burial at Thiais were unprecedented events that reflected the exceptional nature of the crisis, but they were also important components of an emerging and consolidated narrative of the disaster. According to media and political rhetoric, these anonymous victims represented a failure of social solidarity in a culture of entitlement. They achieved public recognition only in death. And in most cases, these were deaths marked by profound indignity, suffered alone in dismal surroundings, while neighbors and public officials vacationed.

This book tells the stories of these victims and the catastrophe that took their lives. It explores three intersecting narratives of the disaster: the official story of the crisis as it unfolded and its aftermath, as presented by the media and the state; the anecdotal lives and deaths of its victims, and the ways in which they illuminate and challenge typical representations of the heat wave; and the scientific understandings of the catastrophe and its management. One of the book's major contentions is that there are significant and even dangerous limitations to official representations of the disaster. A counterhistory of the disaster—a story of the heat wave from the "bottom up," so to speak—provides an important means of contesting those narratives and signaling their incompleteness. The book therefore sets the informal, particular, and often-anecdotal knowledge generated in the collection of these victims' histories of life and death in tension with the globalized, aggregative knowledge of the disaster contained in official reports, media treatments, and scientific analyses. By drawing on these stories it argues that the "forgetting" of those who died during the catastrophe was anything but accidental: that instead, a range of historical factors produced and conditioned the vulnerabilities that marked the victims' lives and deaths. The development of the modern urban landscape in Paris, the perpetuation of social inequalities within its limits, and the political marginalization of the elderly and other vulnerable populations in the course of the last century have produced an unequal burden of risk, which the heat wave revealed with deadly force. From the ways in which the media and the state managed the heat disaster at its outset to the counting of its dead in its aftermath, these accounts of the heat wave point to the processes that pushed a population to the margins of citizenship and public visibility.

The story of these bodies is a complex one, far more difficult to grasp than the select keywords of isolation, abandonment, alienation, and death might suggest. To learn more about the victims of this disaster, I spent months collecting their stories and researching the history of the heat wave in archives and libraries and through conversations with government officials and epidemiologists. What I learned from researching their backgrounds told me less about the victims themselves than about their social relationships and their physical surroundings, and less about the heat wave than about collective memory of the disaster and the populations it most gravely endangered. The stories of these victims are limited in what they can tell us: these individuals represent a tenth of those who died in Paris, and not quite 1 percent of those who died in total in France. The circumstances of their deaths and lives in isolation also indicate that information about this group is not ultimately generalizable to the broader population. Finally, as I argue below, collecting these stories several years after the disaster introduces problems of its own, as the immediacy of the disaster fades along with informants' memories of the situations of the decedents. But there is still essential value in these stories. Not only do they illuminate the aggregate picture of vulnerability that emerges from official reports by putting a human face on the disaster, but they also indicate some of the important limitations of the official narratives. They also tell us a great deal about the social imagination of the disaster and how institutional knowledge has shaped public memory.

In the course of this research, I became fascinated by the ways in which the stories of the forgotten intersected with this institutional knowledge, and how they open a window on the tragedy's multiple social dimensions. The stories of Marie, Pedro, Marcelle, Paulette, Minh, and many others cast a powerful spotlight on the experiences of aging in a changing society, and the all-too-easy possibility of falling through the safety nets of an extensive welfare state. They provide a means of interrogating the conditions of poverty in a society marked by tremendous wealth, and of marginalization in a republican polity. Perhaps most important, they call into question what it means to assess risk, to promote resilience, and to count the dead.

The heat wave of 2003 more closely resembled an epidemic than another natural disaster. Where the Indian Ocean tsunami in 2004, Hurricane Katrina on the Gulf Coast of the United States in 2005, and the Japanese earthquake of 2011 all struck with dramatic, immediate force, the heat wave in Europe appeared as a creeping catastrophe marked by a body here, two bodies there, and only well after its inception a sudden explosion in mortality. Part of this story relates to the nature of heat waves as disasters. Unlike hurricanes or tornadoes, which stop all normal activity in their paths, heat waves appear primarily as nuisances. People will travel into a heat wave on vacation, but not into a flood zone or an earthquake epicenter. Work and recreation go on as planned. A tsunami or a flood drowns its victims nearly instantly; an earthquake's victims are crushed in rubble within seconds or minutes. A heat wave takes days to kill, as bodies slowly deplete their stores of water and as sufficient heat accumulates to raise the core temperature to deadly levels.

Certain populations—in particular, the addicted, the elderly, the sick, and the desperately poor—are at especially high risk for dying in heat waves for a range of biological and other reasons, which distinguishes them somewhat from other disaster victims. The elderly and those taking neuroleptic drugs (typically used to treat schizophrenia and other psychotic disorders) suffer dehydration and heat stroke much more readily than the general population. Aging and drugs break down normal channels of sensory input, so that messages of dehydration do not translate rapidly enough into the sensation of thirst for the body to rectify the imbalance provoked by extreme heat. Alcoholism and other addictions can dramatically magnify dehydration and further cloud sensory input, speeding up these imbalances. Biological risk aside, these conditions—aging, psychosis, drug dependency—often intersect with poverty and social isolation, which themselves aggravate health risk and mortality. Death among these populations produces little shock, as it is already so prevalent. Where a healthy cocker spaniel or a small child who dies while locked in an overheated car is clearly a victim of hyperthermia, who is to say whether a ninety-four-year-old woman or a malnourished, HIV-positive, cross-addicted homeless man died from the heat or from some other cause—old age or Alzheimer's, overdose or AIDS—even if the weather is stifling?

The death of Zoltan, a sixty-three-year-old homeless man who died at the Hôpital Européen Georges Pompidou in Paris's fifteenth arrondissement on 5 August 2003 as the temperature outside soared to nearly 100°F (38°C), thus attracted little official attention. Nor did health officials take note of Patricia's death at age forty-three a few days later, when a desk clerk in a run-down residential hotel in a decaying neighborhood in Paris's eighteenth arrondissement discovered her body against the door: she was apparently trying to open the door when she collapsed. The same went for Françoise, a forty-three-year-old heroin addict and alcoholic, in a squat in an abandoned ramshackle building in Paris's twentieth arrondissement, where she lay dead for over a day until others found her; and for Claude, a homeless fifty-three-year-old man, who died in the street directly in front of one of Paris's coolest environments, a frozen-food outlet.

Although disparate in location, age, and social origins, they had much in common. They had no family to speak of: Zoltan's only relatives lived in Hungary; Patricia, to all appearances mentally ill, had left her family years before without a trace; Françoise had once been married, but lived primarily with her addictions. Their neighbors recognized them, knew a bit about them, but did not save them from their fates. The French welfare state did little to stop their falls. Health officials and the media remained unaware of these deaths, if only because of the ordinariness that surrounded the vanishings.

It is only when these deaths appeared en masse that they demanded public attention. Yet a precise accounting of rising mortality is difficult to achieve. Despite a mounting death toll France's public health system was poorly equipped in August 2003 to recognize the growing catastrophe until the numbers became staggering. Official channels for mortality reporting are slow and cumbersome in France, as in most industrialized countries. It is therefore anecdotal, localized increases in mortality, noticed by individual practitioners, that are far likelier to draw attention than aggregate trends, which can take months to appear. When the story of the heat wave's shocking mortality broke on 10 August, it was thus the nation's public funeral directors and its emergency room personnel—neither of whom have any official sentinel reporting function—who first sounded the alarm.[4] Overwhelmed by the bodies of the dying and the dead, they finally drew media and government attention to a crisis that had been brewing since the arrival of the weather system. For the next three weeks, the killer heat wave was a lead story, buried along with its anonymous victims only in early September, and resurrected repeatedly throughout the following months as official reports came into circulation.

The sudden realization of the unfolding catastrophe forced an instant reimagining of the heat wave. Journalists, government officials, and epidemiologists began to tell a story of extremity and exception. Yet despite these efforts to write off the heat wave as a "natural" disaster—one that was as unmanageable as it was unpredictable, and therefore out of the state's hands—other components of the catastrophe immediately generated a political crisis in France unmatched elsewhere in Europe. The fact that these deaths occurred during the first two weeks of August focused critical attention on the culture of the August vacation in particular: as the young and well-off headed south for holidays at Mediterranean beaches, went the typical story, they left their poor, isolated grandmothers to die horrid deaths from heat stroke and dehydration in their apartments in Paris and Lyon—deaths that a phone call or a visit might easily have prevented. The revelation of so many unclaimed bodies in France's cities reinforced this notion. While deaths in isolation are a fact of modern life, the sheer numbers of these unclaimed bodies underscored the gravity of the catastrophe. The idea that abandonment on such a scale was possible in a nation founded on the notion of fraternity and inclusive citizenship encouraged the story lines of shame and selfishness, entitlement and inequality, and indulgence at the expense of solidarity that already characterized both media and political responses to the catastrophe, to the extent that the "forgotten" of the heat wave became the public face of the disaster.

According to some estimates, heat mortality in 2003 was nearly as high or higher in some countries neighboring France. One study—the only inquiry that assessed death rates in sixteen countries according to the same standard—put Spain's excess mortality during the summer of 2003 at 15,090, and Italy's at 20,089, versus France's 19,490.[5] The factors that influenced high mortality were also similar throughout Europe. Elderly populations—the highest risk group for death by heat stroke and dehydration—are on the rise throughout Europe as a result of low fertility rates and increased life expectancy. French women, for example, can now expect to live to over eighty-four, while women in Spain and Italy can expect to live to eighty-three, compared with eighty in the United States.[6] For much of the twentieth century, France had the oldest population in the world, although now Spain, Italy, and Germany have surpassed it. These countries also have some of the highest old-age dependency ratios in the world—that is, the ratio of those of retirement age to those in the working-age population, which exceeds 35 percent in some regions.[7]

Yet the intense heat disaster of the summer of 2003, despite claiming an estimated 70,000 lives throughout Europe, remains most indelibly associated with France. To some extent this is a result of the political and media crises the heat wave generated in France. As the former health minister Jean-François Mattei (who was widely criticized for his handling of the catastrophe) told me in 2011, the media irreversibly shaped the phenomenon from the outset. The French government, he argued, faced the disaster's consequences immediately, releasing its official mortality figures in September 2003, just weeks after the heat had subsided. Press coverage and politicization during both the disaster and its aftermath were relentless, with hearings in the National Assembly taking place several months later in which dozens of state officials and other actors were called to testify. By contrast, mortality estimates varied widely for Spain and Italy until years after the disaster had passed. In 2003, both countries claimed only a handful of deaths, but revised their numbers upward dramatically in 2005: to some 13,000 in Spain and some 20,000 in Italy.[8] Ironically, both the competency and the relative honesty of the French health ministry in quickly assessing the full measure of the heat wave's toll opened the state to a harsh political criticism from which its neighbors were somewhat insulated. In Mattei's view, this was the fault of a press that covered high mortality in France incessantly, but failed to reassess the French experience in light of revelations of equally or more devastating figures in neighboring countries two years later.

There is some truth to the suggestion that the crisis was more mediatized in France than in Spain or Italy. For example, a media search—an admittedly imperfect metric—indicates that where the Italian dailies *La Reppublica* and the *Corriere della Serra* each ran about 50 stories on the heat wave between 1 August and 30 September 2003, and the Spanish daily *El País* ran 160, the French paper *Le Monde* ran 335 stories on the heat wave in the same period.[9] Likewise, when the Italian national statistics office released new figures indicating far higher mortality than originally suspected during 2003, media attention was slight in France: *Le Monde* ignored the report, and *Le Figaro* ran just one short article.[10] Yet the temporal distribution of mortality is at least as important as these figures. While other European countries experienced what one study has called "minor mortality crises" that occured "almost unnoticed" throughout the summer (albeit with significant elevations in mortality in June and July), the disaster was highly compressed in France, with nearly three-fourths of its excess mortality for the summer experienced in just two weeks.[11] In other words, other European countries such as Italy, Spain, Portugal, and Luxembourg experienced higher than usual death rates through the summer of 2003, leading to a staggering cumulative figure; France, by contrast, experienced its excess mortality nearly all at once, leading to a catastrophic explosion in the death rate in a brief period, making that country a significant outlier.

Yet this book is concerned less with numbers and more with the particularities of the French experience of the disaster. That is, its focus is not on whether the disaster was more acutely felt in France than in other European countries, but instead on the social fault lines the disaster revealed and the historical factors that shaped its course. A growing social science literature has indicated the many ways that disasters illuminate their social contexts. In some cases—the Bhopal explosion, the Indian Ocean tsunami, the September 11 attacks on the World Trade Center and the Pentagon—the state of disaster is acute, generating an immediate and pervasive awareness of the crisis. They represent clear departures from the ordinary, and initiate conditions of near-total disorder: public spaces become shelters, hospitals become triage wards, martial law replaces constitutional protections. Human perception of such disasters has shifted dramatically in the modern era. Whereas the Western world saw the Lisbon earthquake of 1755 as an earthly instantiation of "evil," we now explain such phenomena naturalistically, reserving normative judgments for more clearly fabricated catastrophes such as terrorist attacks or the bombings of Hiroshima and Nagasaki.[12]

Yet the social, economic, and political contexts of disasters remain critical to their interpretation. Disasters are by definition social events, considered in light of their human impact, assessed in measures such as insured losses and mortality. Heat waves, droughts, floods, famines, and tsunamis originate in natural phenomena, but we interpret them though their effects on the local human environments with which they collide. When a hurricane is moving across the sea, it is a threat; when it strikes a community, it is a disaster. The lethality of disasters depends largely on the environment in which they develop. They therefore illustrate the permeability of the boundaries between the natural and the social. Likewise, disasters' effects have important social and political components: droughts, for example, lead to famine only in environments where relief mechanisms fail and when market effects draw food away from those who need it the most. As Mike Davis and Amartya Sen have argued, risk for famine depends more on political than meteorological conditions. Economic factors are also central to the making of disaster: a seismic event such as the Indian Ocean tsunami wreaked its most profound damage in areas where the natural protection afforded by coral reefs and mangroves had been depleted through rampant development and aquaculture, while in the United States, Hurricane Katrina demonstrated with devastating clarity the correlations among vulnerability, social class, and race in New Orleans.[13]

Yet heat waves point with particular acuity to the problematic coupling of human and natural systems. Like all disasters, they have a distinct agency. As the sociologist Eric Klinenberg illustrates in his brilliant analysis of the Chicago heat wave of 1995, factors such as economic inequality, social isolation, and community fragmentation powerfully influence a given population's vulnerability to high temperatures. Klinenberg describes patterns of isolation, fears of urban crime, and ineffective public responses that led heat to kill hundreds of elderly African Americans in the city, but very few whites in the suburbs.[14] Such events reveal human vulnerabilities to extreme crisis. Yet they even more clearly reveal patterns of marginalization that remain hidden in conditions of relative normalcy. As disastrous events disproportionately affect the *most* vulnerable members of a given society, they are assimilated into the fabric of human experience as acute incidences of chronic suffering, and cannot be divorced from the violence of everyday life in at-risk communities.[15] Disasters thus function somewhat like epidemics, which the historian Charles Rosenberg famously described as a social "sampling device." They establish clearly the ways in which human suffering is often more a consequence of social divisions, economic policy, and the failure of political will than of the arbitrary powers of climate.[16]

Drawing on recent work in the study of disaster by geographers, sociologists, anthropologists, and historians, I define vulnerable populations as those that are simultaneously subject to inordinate environmental hazards, with reduced capacity for response, and minimal or nonexistent protection. As a recent panel describes the problem, risk is a product not only of a proximate hazard, but also of the extent of one's exposure to that hazard, other factors that influence vulnerability to the hazard, and the potential consequences of the hazard.[17] These concepts encapsulate the major problematics of the heat wave, which revealed scandalous vulnerabilities in a number of French systems, including emergency medicine, public health responses, government leadership, and mortality reporting.

The book thus merges with an expanding literature on disaster, risk, and vulnerability, while pushing that literature in new directions. My account demonstrates how the heat wave represents an example of disaster marked by a meteorological extreme, but also reveals a high degree of so-called normal risk inherent in the economic inequality and disintegration of community solidarity in contemporary France.[18] It is tempting to frame the 2003 heat wave as characteristic of the "risk society" marked by intersections of ecological disaster, political miscalculation, and the failures of technology and expertise; or in light of the sociologist Charles Perrow's concept of the "normal accident," in which complexity and tight coupling entail hidden vulnerabilities that predispose a technological system—be it a nuclear power plant or an urban society—to unforeseen catastrophe.[19] According to these models, a number of critical factors and interdependencies operated to place French society at risk. Only a sustained heat wave striking a rapidly aging society, living for the most part in cities unaccustomed to high heat and therefore built accordingly, with little air conditioning, could have placed such a large population at risk; only a heat wave striking in August, with most government officials, physicians, emergency services personnel, and families on vacation could have allowed such a high death toll.[20]

Unforeseeability, human error, poor decision-making, and incompetent organizational response played important roles in the heat wave disaster. Yet where the sociological literature on disaster has drawn critical attention to the making of vulnerability, the specific nature of the unequally shared mortality among French populations during the heat wave calls for a model that is more sensitive to long-term historical trends that have shaped risk in the contemporary urban environment.[21] I focus on the ways in which a particular coupling of human and natural systems influenced the disaster's uneven effects on the population.[22] While the heat wave was an unprecedented and extreme hazard, specific historical and social developments—including the evolution of architecture and land use and the marginalization of the elderly and the poor—made certain populations particularly vulnerable to the heat and heightened their exposure to the disaster, producing horrific consequences.

Although much of the research on which this book is founded comes from fieldwork, I write not from the perspective of a sociologist or an anthropologist, but instead as a social historian of public health in the urban environment. This book engages two types of historical inquiry. The focus of the first is contemporary history. Considering how recently the disaster it investigates took place, the book is grounded in the present. Those who study the ancient past must make use of different archives than those that build histories of the modern period: their authors must draw on archaeological evidence rather than ministerial documents, for example. The same goes for the history of the very recent past. The book thus draws on appropriate methods for investigating the present: interviews and observations, close readings of the press, careful analysis of visual media, and research in archives and libraries. It borrows heavily from recent literature in the social sciences, in particular, from work in medical anthropology and the social study of disaster.

Yet its objective is also to establish a rich historical context for the specific vulnerabilities that marked urban France for devastation during the 2003 heat wave. It therefore also draws on another mode of historical inquiry, the history of the present. Borrowing from Michel Foucault's framing of such a practice in a series of works as well as the recently established journal *History of the Present*, this second mode constitutes a genealogy of the now, a careful exploration of the historical record that seeks critical epistemic fractures or ruptures that produce the conditions that contribute to the shaping of the present.[23] The heat wave revealed important fissures in the French social fabric that have colored everyday life for many citizens, and were produced not in August 2003 but over the decades that preceded the disaster. These fissures were not inevitable, but were instead linked to critical moments and contingencies in the past. As the environmental historian Ted Steinberg has argued, a historical perspective provides a critical analytical frame for understanding the social dimension of disaster.[24] Historians' skills allow them to uncover trends in the making over the years and even decades that precede disasters as a means of signaling what went wrong, and when. Although journalists, sociologists, and anthropologists often pay close attention to their stories' backgrounds, their main concern is (rightly) the present. Historians, by contrast, can highlight long-term developments and cultural tendencies that contribute profoundly to the production of local human environments and the risks associated with them.

This book explores two intersecting themes that the heat wave cast in high relief. The first is the social and political marginalization of large populations in the contemporary period; the second is their resulting invisibility. Many factors that predisposed French society to the excess mortality experienced in August 2003 have been in the making since at least the interwar period. These historical processes, sometimes decades in the making, helped to structure risk and shape the outcomes of the disaster. The late twentieth century has produced in France a landscape of vulnerability in which specific demographic transformations, organizations of the built environment, and responses to social change coincided with particular ideas of the rights and responsibilities of citizenship to contribute to the making of disaster. Many of those who died during the heat wave lived on the edges of society. Whether by virtue of their destitution, their advanced age, their physical or mental disability—or combinations of all of these things—they found themselves at the limits of humanity. This book draws on several interpretive frames to explore some of the ways in which the experience of the heat wave disaster in France sheds light on important social and political developments in that country in the past several decades.

One of these frames derives from the centrality of citizenship to modern French history. As I note in chapter 1, the heat wave and its management elicited a vigorous debate about the relationship between the individual and the state. Where many on the political left faulted the center-right government of President Jacques Chirac and Prime Minister Jean-Pierre Raffarin for failing to protect France's most vulnerable populations, many on the right blamed a collapse in social solidarity and declining family values for the crisis's unprecedented death toll. An emerging literature on health, citizenship, and vulnerability poses useful questions for the French experience of the heat wave. Where a traditional conception of citizenship involves access to the ballot box—and in particular, the efforts of women, religious minorities, colonized populations, migrants, and other populations excluded from the franchise to obtain access to the political process and basic civil rights—social scientists have expanded the meanings of citizenship to include far more than eligibility for political participation and civic equality. As early as 1949, T. H. Marshall argued that contemporary citizenship (at least in industrialized countries) included assumptions of basic economic and social protections; four decades later, the sociologist Bryan Turner considered citizenship a marker for, or even constitutive of, a broader social integration and solidarity.[25] In France in particular, the issue of citizenship has involved far more than strictly legal questions.[26] As Joan Wallach Scott has argued persuasively, debates over republican universalism in the past two decades have profoundly informed discussions of legal and social citizenship. The founding ideas of the nation in the revolutionary period defined citizenship through recourse to universalist principles: any citizen could represent the nation only because citizens were capable of abstraction from particular social attributes. This was not so much a denial of particular differences, but merely a recognition that those differences belonged to private rather than public life. Yet as Scott argues, some differences—for much of French history, sexual difference, and in the late twentieth century, religious or ethnic difference—have proved irreducible and incapable of abstraction, thus erecting important barriers to the admission of women and immigrant populations to full citizenship.[27]

This book argues that certain other social categories have remained incapable of abstraction and representability. The elderly poor (and in particular, elderly poor women), the homeless, the disabled, and the mentally ill—all of whom are likelier to die during heat waves than the general population—also resist easy assimilation into the model of republican citizenship. This relative exclusion has had real consequences for political representation and full access to the rights of social citizenship, and has also resulted in the increased vulnerability and even invisibility of the elderly, the poor, the disabled, and other groups at high risk of dying during heat waves. Their marginalization derives partially from a perceived ineradicable difference inherent in these populations (as Scott describes was operative in the exclusion of women from political rights), but also, I argue, from a differential valuation of life that is both a determinant and consequence of social and political investment in some populations at the expense of others. Moreover, these two mechanisms work in tandem rather than as separate techniques of exclusion.

As the sociologist Robert Castel has argued, one of the social mechanisms that pushes some to the margins while overvaluing others is a process of what he calls "disaffiliation." Castel describes the emergence in the postwar period of a *société salariale* in which the welfare state has become bound to employment status through a series of public and market forces. As some become distanced from the labor market—through unemployment, age, or disability—they become gradually disaffiliated from state and society, and therefore from full citizenship.[28]

This is clearly a constitutive factor in the marginalization of elderly, disabled, and homeless populations in France. But such social and political exclusion is also a product of the historical development of the modes of objectification that Michel Foucault studied extensively. In much of his work, Foucault documents the operation of dividing practices. These are techniques of domination that define and separate marginal populations according to varied historical exigencies: lepers from the sound of body, the mad from the sane, the sick from the healthy, the sexually deviant from the mainstream. Historically contingent social, scientific, and political discourses give rise to broad epistemic tendencies that come to mark a society and period. Foucault's works on the histories of insanity, of medical pathology, and of sexuality all point to epistemological shifts in the notion of the human as defined through the not-quite human or the otherly human, something walled off emotionally, intellectually, and often physically as a means of protecting the "normal" from the incipient threats of the "pathological." Foucault's outline of the concept of biopower is perhaps most relevant to this project. Beginning in his lectures at the Collège de France in the late 1970s, and through his three-volume *History of Sexuality*, Foucault describes the operation of what he calls a "biopolitics of populations." Emerging alongside a science of population in the late eighteenth century—the state's gathering of vital statistics, demography, and by the early nineteenth century, a nascent hygienism that aimed at the improvement of public health—biopolitics constituted the form through which state institutions could influence the nature of the population. Where the sovereign (in the figure of the monarch) once had the power to kill those who threatened royal authority, a new form of authority in the aftermath of the Revolution offered the state the capacity to direct life. Bureaucratic centralization and coordination meant that new state and substate agencies could direct resources toward increasing the birth rate, encouraging healthy behaviors, and investing in the improvement of the population. In Foucault's famous formulation, where once the monarch had the power "to *take* life or *let* live," now the state and its institutions had the power to "*foster* life or *disallow* it to the point of death": to direct the lives of some toward improvement, while allowing others to die through neglect.[29]

A number of subsequent critics have developed the theory of biopower in fascinating ways. The Italian philosopher Giorgio Agamben in particular has modified some of the central tenets of Foucault's writings on biopower to explore the subtle mechanisms that mark populations for death. Where Foucault draws a temporal break between a productive biopower that encourages the direction of life and a destructive sovereignty that characterized the ancien régime, Agamben sees these forms of power over life and death as linked and inherent to the human condition. Agamben takes up Hannah Arendt's observation that classical Greek contains two expressions that mean "life": *bios*, or biographical, political life marked by full humanity, and *zoë*, or simple animal existence. He then develops this concept through reference to an obscure juridical figure in ancient Rome, the *homo sacer*, or the figure who can be killed but not sacrificed: that is, a figure who through the violation of certain laws has become a bestial outcast endowed with the political status of an animal, a status Agamben calls "bare life." From there Agamben draws a line through the medieval European ban, or the process by which bandits were exiled outside city walls—to an existence, that is, of noncitizenship, outside of collective political life—to the politics of Nazi Germany. These domains had in common the capacity, through rhetoric and law, to reduce *bios* to *zoë*: to collapse human life into an animalized existence. Only the prior biopolitical reduction of Jews to the status of bare life, Agamben argues, allowed for their mass extermination. In the present, Agamben finds the figure of the refugee to be the new avatar of bare life, someone who is explicitly outside the domain of citizenship and who exists in a near-permanent status at the threshold of death.[30]

Some have criticized Agamben for his contention that the Holocaust and other genocides can serve as a general frame for imagining biopolitics in the twentieth century. Nikolas Rose, for example, sees this as going a step too far, and placing too much agency in the hands of the state: although life may be "subject to a judgment of worth," the state itself does not organize the death of its citizens. "To let die is not to make die," Rose argues.[31] And the Finnish political philosopher Mika Ojakangas argues aptly that the central avatar of modern biopolitics is not the extermination of the Jews, but instead middle-class Scandinavian social democracy.[32] But despite these useful criticisms, a number of scholars have drawn on Agamben's frame of bare life and sovereignty to refer to other contemporary forms of marginalization and violence. The anthropologists Peter Redfield and João Biehl have found this concept useful for explaining the ethical dilemmas of humanitarian aid and the social abandonment of disenfranchised populations, respectively.[33] For Redfield, the global engagement of aid workers has supplanted the state in many developing countries, preserving populations at a bare level of existence through a humanitarian minimalism. For Biehl, it is precisely the retreat of the state in its privileging of the market that has engendered the social abandonment—the reduction to bare life—of a growing population of the mentally ill, the disabled, and the paperless, who find themselves caught between legal, charitable, and public health institutions. While the state may not be "willing" its populations into sickness and death, to recall Rose's term, its neglect of those most in need in the face of market forces amounts to a powerful form of dehumanization—a phenomenon that is operative in totalitarian societies and, as this project indicates, in liberal democracies alike.

Judith Butler provides a useful elaboration on these processes of degradation, dehumanization, and suffering that offers both a corrective to Agamben and an expansion of the possibilities of this body of thought. For Agamben, bare life is a biological essence that is universal to humanity. Yet the dehumanization of populations happens unevenly, and Agamben fails to describe how the exercise of power toward this end unfolds in specific instances and among particular populations. Butler argues that the "targeting" and "management" of certain populations and the "derealizing" of their humanity are rooted in historical specificity and play out along lines of race, ethnicity, religion, sex, or other social categories. In a series of essays on the war on terror and the Iraq war published in 2004 and 2009, Butler argues that certain rhetorical and visual frames create a differential valuation of life, endowing some lives with "value and dignity" and "foreclose[ing] responsiveness" in others.[34] The deaths of some thus constitute enormous tragedies that generate powerful commemorations—here, Butler writes of Daniel Pearl, the Israeli-American journalist and young father whom al-Qaeda terrorists kidnapped and beheaded in 2002. Other deaths, by contrast—those caused by drone strikes and counterinsurgency in Afghanistan and Iraq—become anonymous and unmeaningful through processes of representation that render some deaths assimilable and others foreign to our experience.[35]

This book borrows from each of these concepts to argue that parallel processes of degradation have enhanced the suffering of some of France's most vulnerable subjects, pushing them to the margins of citizenship and even humanity. The media's representations of heat wave victims in general, and the forgotten victims buried at the Thiais cemetery in particular, defamiliarized them. It made them incapable of representation in universalist terms, and in so doing, removed them from full humanity. As a consequence their deaths became, to use Butler's term, "ungrievable," or incapable of provoking outrage or even anything more than a modicum of sympathy: their lives left "no public trace," or at best, "only a partial, mangled, and enigmatic trace."[36] Certain types of storytelling underscored this marginalization: media depictions of the victims, political and scientific discourse about the management of the disaster and the factors that exacerbated risk for dying during the heat wave, and the narratives that informants shared with me in the course of my research all point to a broad assumption that the tragedy of the heat wave was not a global one, but that it instead preyed on particular and idiosyncratic populations that could not be generalized to the French population as a whole. I also argue that historical processes contributed significantly to the generation of these vulnerabilities. In the past century, urban development and market forces have combined to make the lives (as well as the deaths) of some populations almost invisible, while political discourse and public policy have pushed the elderly to the margins of society. Finally, the science of counting the dead during the disaster implicitly valued some lives more than others through different strategies for measuring the impact of the heat wave on mortality, a process that has had important repercussions for policies designed to protect the population from future disasters. All of these factors had the result—through rhetoric and practice—of stripping away the full citizenship and humanity of certain sectors of the French population, and in particular, those who were most likely to die during the catastrophe.

The predicament of France's most vulnerable populations during the heat wave is thus linked to a fundamental economic and social degradation of life. I use the term "degradation" intentionally because of its connotations of entropy and decaying integrity. All French citizens are born with political legitimacy and legibility, with the state as the guarantor of human rights: in Agamben's words, it is "the pure fact of birth that appears . . . as the source and bearer of rights."[37] Yet where Agamben sees juridical power as the origin of a state of exception that excludes some from the rights of citizenship, I argue that, instead, it is social and rhetorical forces that cause some to lose this status over time and to collapse into conditions of vulnerability. As Foucault's student François Ewald argues in his magisterial discussion of social protection, the welfare state represented a formalization of the state's protection of life and its investment in the family as a foundation of the republic and the social, so that private matters such as birth, marriage, and kinship became public institutions through the relationship of the individual to the state.[38] As these bonds are mutually constitutive, so is their fraying: a rhetorical dehumanization exists in tandem with economic disaffiliation and the fracturing of social protection.

The production of vulnerability is also a consequence of widespread processes of forgetting, which constitutes another major theme in this book. The heat wave primarily struck down those whom economic development and market liberalization have left behind. The heat wave's death toll offered a stark reminder of how populations living on the edge of citizenship and humanity remain at the heart of the city, and in significant numbers. Yet in everyday circumstances, the elderly poor, the homeless, and the disabled occupy a small place in the consciousness of most.

Moreover, this forgetting is far from benign neglect. In his introductory essay to the edited volume *Agnotology*, the historian Robert Proctor points to three specific kinds of ignorance that operate in ways that are central to modern scientific thought and practice: ignorance as a "native state," from which scientific inquiry emerges; ignorance as a "lost realm, or selective choice," which is a result of the decisions to advance some inquiries while overlooking others; and ignorance as an "active" or "strategic ploy," which involves the deliberate production of uncertainty, as in the politicization of knowledge about tobacco, industrial food, and climate.[39] The sort of ignorance or forgetting at play in this project is the second: ignorance as a consequence of selective choice. Both scientific and social knowledge develops as a result of the direction of inquiry and the prioritization of concerns. But the development of certain kinds of knowledge does not merely leave areas of ignorance untouched: it actively expands them. Putting the spotlight on certain problems and concerns diverts that light from other pressing issues, allowing "darkness to grow as fast as the light," to use Proctor's words. As Londa Schiebinger argues in her essay in the same volume, European explorers who sought new materia medica in the West Indies saw abortifacient compounds that West Indian women used as irrelevant to their concerns. Naturalists' decision to invest in the research and collection of other plants and to ignore abortifacients not only left that knowledge fallow, but effectively buried it over time.[40]

There is a close connection here between the sort of ignorance that emerges from social and research prioritization and that which structured the vulnerability of France's forgotten victims of the heat wave. An active investment in some forms of population development, combined with a media and political rhetoric that has historically marginalized others, has produced a widespread invisibility of certain subjects. Other terms that the press used to describe the forgotten victims underscore this invisibility: *les morts dans l'anonymat* and *les corps anonymes* offer useful metaphors for the place of these subjects in the contemporary urban landscape as reservoirs of a lack of knowledge. As this book argues, the heat wave's forgotten victims rapidly became a metonym for the entire population that died during the disaster, implying the widespread marginality of those who died. If, as Schiebinger argues, societies develop certain types of knowledge rather than others—a marker for their investment in certain populations rather than others—then there are clues in those societies and their politics that hint at the complex processes that render some populations invisible, incapable of assimilation into the universalist representability that constitutes a prerequisite for full citizenship, and subject to a revocation of the social contract.[41]

The chapters in this book explore the intersections of vulnerability, invisibility, and marginalization by examining three different types of narratives that appeared repeatedly during the disaster and its aftermath. The first of these is a narrative of vulnerability. Who was at risk during the crisis, and why? What do the deaths of the elderly, the marginal, the addicted, tell us about French society at the beginning of the twenty-first century, and how did the constant reiteration of victims' vulnerability itself contribute to their marginalization? The second, closely related theme examines marginalization in the urban built environment. Who were the anonymous victims of the heat wave, and what are their social histories—literally, the stories of their integration into society, as told by neighbors, relatives, and the media? What do these stories tell us about life and death on the margins in the modern West? The third is a narrative of explanatory power: the capacity of epidemiology and demography to encapsulate the heat wave's revelations about society, and to frame the story of the disaster. How did the state manage the crisis as it unfolded and in its aftermath? What are the implications of the sciences of population, vulnerability, and mortality for relationships between the individual and the state in a moment of crisis?

Chapter 1 relates a chronology of the disaster. Drawing on media accounts and the reports of a number of state agencies as well as hearings in the French National Assembly, it discusses the heat wave's meteorological origins and effects, describes its impact on the French population, and presents an outline of individual and government responses to the crisis. Its scope is the seven weeks between the onset of the heat wave and the publication of the official report of the death tally, which established the disaster's excess mortality at 14,802 victims. I uncover the day-to-day conditions of the heat wave, and emphasize the gradual realization of a catastrophe in the making. The chapter has several goals. One is to orient the reader to the disaster's scale and its wider political and social context. Another is to detail the dysfunction that characterized the state's response to the disaster. And a final goal is to indicate the ways in which media, political, and scientific accounts of the disaster—as it unfolded and in its immediate aftermath—portrayed the heat wave's victims in a light that exacerbated their vulnerability through a relentless emphasis on their marginality.

Chapter 2 builds on this last aspect of the first chapter. It delves into the story of the forgotten, the hundred-odd bodies abandoned to public burial at Thiais. Here I describe the demographic characteristics of the group, and highlight the ways in which the forgotten are both representative of and an important departure from the aggregate portrait of the "typical" heat wave victim as presented in state reports. The chapter tells the story of the forgotten from a media and political perspective, interpreting closely the dozens of stories produced about the group in the daily press, magazines, television news, documentary film, and books. It travels from the cemetery at Thiais to nearly every section of Paris, from luxurious buildings in the city center to the sidewalks of decaying neighborhoods, detailing the methods I used to track down the individual stories of those who frame this book. But it also highlights the powerful intersection of invisibility and vulnerability, of forgetting and marginalization, that the disaster revealed. The stories of the heat wave's forgotten victims provide an indication of how public memory of the disaster has contributed to a widespread imagination of those who died as social isolates who had made their own vulnerability and predetermined their own fates.

Chapter 3 investigates the importance of place to the shaping of the disaster's outcomes, showing how the urban built environment is a technological system capable of breaking down in its own right. I explore the forms of vulnerability built into individual buildings, neighborhoods, and the layout of the city itself. There is a long history of Parisian urbanism and the evolution of ideas about the city as a site of sickness and health, beginning with the first medical geographies of Paris in the mid-nineteenth century and continuing through the present. Drawing on this rich history, I explore the spatial dimensions of vulnerability by investigating patterns of mortality throughout the city both in normal periods and during the heat wave; but I also note a vertical dimension of risk, through the cases of those among the forgotten who died in tiny quarters (sometimes of less than 100 square feet) directly under zinc roofs on the top floors of sometimes luxurious and sometimes decrepit buildings. There are pockets of desperate poverty even in the city's most chic quarters, and official efforts to rectify an illegal housing market that perpetuates inequality and vulnerability for the city's poorest residents have fallen far short of protecting those residents from the vicissitudes of the market.

Of the heat wave's 15,000 victims in France, some 80 percent were over seventy-five years old. In chapter 4 I explore the transformation of ideas about aging in the past century. After interviewing a number of those who lived alongside the elderly victims of the heat wave and consulting a range of archival sources, I develop a historical context for their lives and deaths. There are both real and imagined connections among aging, disability, and poverty, and the ways in which those intersections have shaped political and social discourses about the elderly in the postwar era reveal a history of the biopolitics of aging. Offering a meditation on the relationship between health and citizenship, I chart a course that has led to the systematic social and political marginalization and dehumanization of the elderly in contemporary France. The heat wave, I contend, represents a culmination of these processes in two ways. On the one hand, it forced a collective realization of the social impact of aging in France. On the other, the marginalization of the elderly that so dramatically skewed mortality in their direction is itself an end result of the conceptualization of aging as a social burden and a political problem.

Chapter 5 investigates cases drawn from among those buried at Thiais to explore vulnerability in groups other than the elderly. Calling on the examples of a suicide, a homeless heroin addict, an alcoholic man in his forties, and an AIDS patient among others, it asks the question, who is a disaster victim? Among the most intriguing elements of the field interviews I conducted was a reluctance of informants to ascribe the deaths of those other than the elderly to the heat. Where the neighbors of the elderly considered the deceased to be clear victims of the heat wave, the familiars of other victims emphasized other possible causes of death. By their accounts, it was alcoholism, it was depression, it was HIV, it was homelessness that killed their neighbors: the heat certainly did not help matters, but they should not be considered heat victims. These victims open a discussion of three issues. The first is an engagement with the science of epidemiology and the amassing of vital statistics. The counting of the dead and the attribution of causality are practices that are fraught with uncertainty. While the forgotten victims all figure in the global mortality statistics of the disaster, a closer examination raises serious questions about the cause of many of their deaths. At the same time, the deaths of many of these figures draw the focus away from the elderly as the heat wave's principal victims. My second goal in this chapter is thus an investigation into the other forms of marginality that determine vulnerability in periods of disaster. Policy initiatives focusing on enhancing resilience among the elderly during heat waves will have little effect on victims like these; they remain at high risk. Finally, I examine how aggregate pictures of vulnerability stereotype victimhood. In the aftermath of the heat wave, it became easy to associate vulnerability with advanced age and poverty, but increasingly difficult to ascribe the deaths of nontypical victims to the disaster rather than to another cause.

The book's epilogue focuses on the tensions between the ethical dilemmas that the heat wave forced on the French public and the continued invisibility of vulnerable populations. It begins with a reading of a 2004 novel set during the heat wave, Thierry Jonquet's *Mon Vieux*, which asks the question, should a society invest in the old, or in its young? Should it preserve life at its end, or should it foster a better life for all its citizens? Is the welfare state capable of all of this, or only a part, and if the latter, how shall it prioritize the lives of some over others? The epilogue summarizes the ways in which city and national officials have attempted to come to terms with the revelations of the heat wave of 2003 and to prevent its recurrence. Local and national publicity campaigns have sought to generate awareness of the vulnerability of the elderly in order to mitigate the effects of future heat waves. When the heat struck again in July 2006, a far lower death toll convinced authorities of their methods' success. Yet there are important limitations to the state's programs that are revealed by the experiences of many of the forgotten: the homeless, the mentally ill, the figures readily discerned in the cityscape but who defy easy intervention. The book concludes with a discussion of the increased interest in sustainability in France—a phenomenon hardly mentioned in political and social life before 2003, but a regular fixture in the aftermath of the heat wave and rising concerns about the impact of climate change on human health.

One final note is in order. This project would have been impossible were it not for the work of two authors in particular. It owes a profound debt to Eric Klinenberg's *Heat Wave: A Social Autopsy of Disaster in Chicago*, which details a rich investigation of the social factors that shaped vulnerability during the Chicago heat wave of July 1995, and highlights the city's mismanagement of the disaster. Many of the main themes in this book are borrowed from Klinenberg's work: the place of aging and isolation in a postindustrial society, the meaning of death in anonymity, the importance of place and sociability to the shaping of vulnerability, the role of the media and political decision-making in structuring risk, and the particular agency of heat waves as lethal disasters all figure prominently in Klinenberg's book. Our conclusions are somewhat divergent, but as a function of geography and culture rather than intellectual differences: while heat mortality in Paris in 2003 had much in common with that in Chicago in 1995, significant demographic and cultural variations between the two sites mean that risk had a somewhat different profile in the two sites as well. This book also owes a methodological debt to the anthropologist Jeanne Guillemin's *Anthrax: The Investigation of a Deadly Outbreak*, which explores a devastating 1979 epidemic of inhalation anthrax in Sverdlovsk in the former Soviet Union.[42] My book stems from concerns similar to those of Guillemin: a dissatisfaction with official explanations of a public health crisis and an eagerness to explore the social memory of a disaster. Like this book, Guillemin's begins at the graves of a catastrophe's victims and proceeds with an effort to tell those victims' stories. Our goals are different, but our investigations yield a similar result: they indicate that the stories of individual victims provide an important means of enriching—or at times contesting—official representations of catastrophe.

In borrowing extensively from qualitative social science, this book does not seek to imitate that work, much less to suggest the obsolescence of historical inquiry. But it does seek to push the boundaries of what it means to be a historian. By definition, we work in the past and the present, chasing after the ghosts that we glimpse in archives and libraries, in seminar rooms and in interviews. In writing about the past, we typically search for traces of the dead in the present—the agents of history and their legacy. It is the spectral presence of past populations and landscapes that we seek to detail. Writing a history of the present complicates this project greatly, as it involves both living and dead actors. This book relates a number of stories about the heat wave as a social, human, and natural disaster. But it also aims at describing what happens when one literally knocks on doors and looks for the past's dead in the now. Although I looked for ghosts, what I found was memory; although I looked for the social dimensions of disaster, what I found were the anecdotes that are the building blocks of a constructed past and its uses in the present.

1

Stories, Suffering, and the State: The Heat Wave and Narratives of Disaster

In early August 2003, a warm air mass began to move northeast from the Mediterranean across Spain and into southern France. In the course of the next few days, it crept northward, expanding its reach and eventually stabilizing over western and central Europe. The mass was part of a phenomenon known as an "anticyclone," a large-scale system bound by rotating winds and marked by extremely high pressure that prevents others from displacing or disrupting it. Such systems are common in European summers, in varying degrees of intensity. By 4 August, this anticyclone was firmly implanted, bringing with it a stifling heat with little to no breeze and no precipitation. Much of Europe smoldered under an intense sun for nearly two weeks to follow. The system also intensified concentrations of ground-level ozone. Combined with the overwhelming humidity, the system produced a dizzying cocktail of heat and pollution in urban areas.

While the system encompassed much of Europe, it was particularly intense in France, where it left nearly 15,000 dead in its wake. This chapter plots the chronology of the heat wave disaster, while also presenting the ways the French government and its epidemiological services responded to the crisis in real time. It sketches the dominant narratives that have come to frame the catastrophe in France in its aftermath: those of the media, of the state, and of the epidemiologists who studied the disaster at local and national levels. Finally, it indicates how the state and the media collectively produced a portrait of the heat wave's "typical" victim and outlined a series of natural and social causes responsible for the making of the disaster. This narrative obviated the state's culpability for the mismanagement of the catastrophe, while assigning responsibility for the heat wave's staggering effects to forces beyond the government's control: natural systems, demographic transformations, failures of social solidarity, and the social particularities of the victims themselves. Such a frame—which epidemiological reports on the disaster reinforced, even if inadvertently—ultimately redirected blame from the state to the heat wave's victims by underscoring their marginality. This chapter indicates that this gradual realization of a state of disaster, the recriminations that followed, and an effort to establish control over the story of the heat wave distilled the complexities of the catastrophe into a streamlined narrative: one that incorporated many of the disaster's critical truths, but which also transformed them into a misleadingly simplistic fiction.

A Disaster in Slow Motion

The August heat wave (or *canicule*, in colloquial French) was in some ways more of the same for the French in the summer of 2003. The country had suffered several heat waves already that summer, the first lasting several days in June and two others lasting roughly a week each in July. Moreover, France had suffered a drought since February of that year. Rainfall levels were down by 50 percent in most regions, and by 80 percent in many more. By June the landscape in much of France was dessicated, more closely resembling typical September conditions than those at the beginning of summer. Wildfires are often a summer problem in France, but in 2003 they scorched more than three times the annual average. Media coverage throughout the summer focused on the devastating consequences of the heat and drought for the nation's farmers and others who made their living depending on the weather. Emaciated livestock, parched fields, and blazing forests featured regularly on the evening news.

But for most of the French, the summer heat in June and July was merely an inconvenience. Despite the heat's growing intensity and its effects for farmers, most of the country seemed oblivious, with a few exceptions. The provincial press covered the heat as a human-interest story above all: they showed photos of dogs lazily sleeping off the heat and compared third-trimester pregnancy in such conditions to the stations of the cross. Stifling temperatures in Paris created a boom for café owners, and especially for the ubiquitous bottled-water vendors stationed outside the city's museums and along tourist thoroughfares, many of whom doubled their prices amid the insatiable demand for cool drinks. Fountains in Parisian parks proved a big draw for tourists, and fans and air conditioners sold out quickly.

When Météo-France, the French national weather service, issued a mild press release warning of an intensifying heat wave on 1 August, the heat appeared primarily as an annoyance. The media focused a number of stories on the heat wave, but mostly in terms of its potential effects for the August vacation period. Since the late 1930s, the August vacation has been a sacrosanct social entitlement in France. Beginning with the Popular Front in 1936, the paid vacation—with enticements to tourism such as discounted rail fares and holiday packages—became an inherent component of French cultural identity. Popular Front leaders encouraged such vacations as a means of stimulating the nation's (and in particular the worker's) physical as well as intellectual culture toward the end of producing a total individual. The paid vacation is arguably the movement's signature achievement as well as its chief legacy.[1] As Ellen Furlough has argued, the political securing of paid vacations as a right of citizenship, along with the emergence of a postwar consumer culture, has made tourism a mass phenomenon in France: since the 1980s nearly two-thirds of the French travel during the summer vacation in particular.[2] With most of this travel moving from cities toward beaches, the countryside, and mountain resorts, the result is an enormous outward migration from urban areas the first weekend of August.

In 2003 the beginning of the vacation exodus took place on Saturday, 2 August, just as the temperature began to climb. The news that day featured prominent stories about how the heat intersected with the intense highway traffic that marked the departure of millions on vacation. Most of the stories featured interviews with drivers on the highways complaining about sweltering in their un-air-conditioned cars while stuck in traffic jams; others showed tourists napping in the shade at rest stops en route to their destinations. But the overall atmosphere was cheerful, with the heat presenting a challenge to what some voyagers described as an obligation to travel: "For vacation, you have to do what you have to do!"[3] A few warnings stand out, but not really about the temperatures or about dehydration. One meteorologist reminded motorists to wear their seat belts while traveling; a reporter for the major television network TF1 remarked, "At the beach, the only worries are sunstroke and sunburn." But for the most part, the weather seemed to be a cause for celebration rather than foreboding: a sunny, albeit hot, forecast awaited vacationers at Mediterranean beaches and Alpine campgrounds for days to come. As a meteorologist for France 2, another of France's most popular television networks, announced on the evening of 1 August, "The water's warm. . . . Take advantage of it!"[4] On the same day, Catherine Laborde, a meteorologist for TF1, described the anticyclone as "protecting" or "sheltering" France from clouds and rainfall.[5]

Meanwhile, out of sight of the media, Bodo was dying from the heat in his apartment in the Boulevard de Port-Royal in Paris's fifth arrondissement. Bodo had emigrated from Germany decades earlier, and lived alone in a tiny apartment on the sixth floor directly beneath the building's zinc roof, with western-exposed windows. Bodo had few social contacts; his only known family member was a half brother in Germany, who had met him only once for about a half hour thirty years earlier.[6] Bodo's apartment was a *chambre de bonne* or former servant's quarters, converted into an independent apartment of about ninety square feet. When I visited his apartment in June 2007, the minuscule elevator was stifling, despite a comfortable temperature in the low 20s C (low 70s F) outside; a downstairs neighbor with the same exposure had covered the apartment windows with newspaper to block the blazing afternoon sunlight. In August 2003, his neighbors on the lower floors were all away on vacation, according to one resident. But one of his neighbors on the sixth floor had remained in the city and discovered Bodo's body on 2 August. The neighbor saw that Bodo's door was ajar, and after calling for him, attempted to open the door. Bodo's body was blocking the open door, as if he had fallen to his death while opening the door to seek help.

In the coming days, troubling signs of a disaster in slow motion began to accumulate. The heat and the drought were consistently the top stories in national media. While some commentators refer to heat waves as the "neutron bomb" of natural disasters[7]—crises that kill people while leaving infrastructure largely intact—the French heat wave, combined with the brutal drought, was anomalous, presenting its effects on the landscape and the built environment before it registered on human bodies. Reports in early August focused on the weather's effects on local ecologies and infrastructure, as they had for much of the summer, but with a new degree of urgency. Beginning on 1 August, nearly all domestic news coverage was connected in some way to the heat. Reports of uncontrollable wildfires dominated coverage, followed by stories about excessive ozone pollution fouling the air of France's major cities as a consequence of the heat. Farmers complained of drastic losses and pleaded for government aid to buy water and hay for their livestock.

The problem was national in scale. The volume of the Loire River had shrunk by 20 percent as a consequence of heat and drought, decimating hydroelectric power generation in the region and threatening a number of fisheries. Reduced flow and stifling temperatures also forced officials to shut down several nuclear power stations in order to prevent possible meltdowns.[8] Melting glaciers in the Alps led to unprecedented avalanches and other risks at heavily touristed sites such as Chamonix, where the Aiguille du Midi, a famous climbing site, was closed to mountaineers for the first time in history. At Mont Blanc, thirty-eight alpinists were evacuated when a rockslide linked to melting ice cut off their route to safety.[9] When storms hit areas desiccated by the heat, ensuing mudslides ensured the complete destruction of vulnerable landscapes.[10] By 9 August, many of Paris's city parks were closed because of an "unprecedented phenomenon": drought- and heat-stricken trees were losing their limbs, and unpredictably falling branches posed a danger to the capital's citizens seeking respite from the heat in the city's limited green spaces.[11]

Not all of the news linked to the heat was bad. In the Ardèche and other wine-producing regions, the harvest came two to three weeks early. While the heat meant less fruit overall, it produced greater concentrations of sugar in each grape, resulting in outstanding wine.[12] But devastation to the landscape and to the built environment greatly outweighed the heat's few advantages. Warmer temperatures proved hospitable to new invasive agricultural pests, hitting farmers when they were already on their knees.[13] The heat hit poultry farmers particularly hard: by 8 August, over a million chickens had died from the heat, presenting a public health disaster to those charged with disposing of their bodies.[14] Meanwhile, temperatures hit 120°F (nearly 50°C) at Formula 1 tracks—and far hotter on the asphalt—presenting major dangers to drivers as their tires melted from the heat.[15] On 6 August, TF1 recounted the story of a "nightmare journey": as high temperatures melted catenary wires that supplied electricity to the national rail system, a high-speed train traveling from Paris to Hendaye in the southwest stalled overnight for six hours, stranding passengers in overheated train cars. The coming days brought more such incidents, including swelling and deviation of the rails as a consequence of the heat, with delays stranding thousands during one of the country's busiest travel periods.

While much of the country vacationed, the heat began to raise a few alarms for health authorities. On 4 August, Météo-France issued what amounted to the first national warning with a posting to its website that noted that "every year around the world, heat, perhaps even more than cold, kills." The website offered a number of guidelines for coping with the heat, advising the French to hydrate constantly, to cool off by spritzing the body with water, to cover windows with sheets soaked in water, to dress in light clothing and to eat light foods, and to frequent air-conditioned and other cool environments including green spaces.[16] The next day, the department of Morbihan in northwestern France reported the country's first official deaths linked to the heat wave. A local health worker noted that three unrelated middle-aged men—a cannery worker, a mason, and a municipal employee aged thirty-five, forty-five, and fifty-six, respectively—died of heat stroke. According to their treating physicians, all three were obese alcoholics, two with psychiatric histories and the third with a history of hypertension. Two died with body temperatures exceeding 107°F; the third with a temperature exceeding 104°F.

On 6 August, the official signaled the deaths to the Direction Générale de la Santé (DGS)—a division of the Ministry of Health charged with the administration of public health policy, health surveillance, and health security—which in turn directed the local official to the recently established Institut de Veille Sanitaire (InVS). The InVS, modeled in part on the U.S. Centers for Disease Control and Prevention (CDC) in Atlanta, had been charged since its creation in 1998 with epidemiological surveillance and alerts throughout the country. Dr. Loïc Josseran, a physician working with the InVS, investigated the deaths the next day, concluding that "these deaths are certainly connected with the excessive heat of these past few days." His report highlighted the men's medical histories, and ended by asking whether a "national census of these deaths" might make sense "in these circumstances."[17]

At the DGS, little was happening at this point. Lucien Abenhaïm, the head of the directorate who reported to the health minister, Jean-François Mattei, was on vacation; his replacement supervisor, William Dab, an English-trained epidemiologist, was about to depart. On 6 August, as he left on vacation, Dab sent an e-mail to the new acting head, Yves Coquin, warning that epidemiological history suggested that "we might anticipate some excess mortality linked to the heat wave. It would be useful for the DGS to prepare a press release reminding about some basic precautions, notably for newborns and the elderly. There are numerous studies on the health impact of heat waves."[18]

By every indication, Coquin took the warning seriously. In the next two days, correspondence within the DGS and between the DGS and InVS suggests that he worked assiduously on producing an effective press release, and also pleaded with the InVS to begin a rigorous investigation of the links between heat, morbidity, and mortality in the country. Yet there were distractions that appeared more urgent. An outbreak of Legionnaires' disease in Montpellier on 7–8 August took the spotlight away from the heat wave for several days, and high concentrations of ozone pollution linked to the anticyclone threatened to produce a number of health crises for asthmatic and other pulmonary patients throughout the country.

Moreover, the deaths in Morbihan were easy to explain away. Yes, they were clearly heat-related, but given the men's medical histories, they were unsurprising given the intensity of the weather. The same went for a death reported in Paris on 7 August of a fifty-six-year-old man on his way home from work. Such cases appeared isolated, if not obvious examples of the sort of mortality one might expect during an unprecedented heat wave. When several more reports came into the agency on Friday, 8 August, they were even clearer cases of the types of deaths to be expected during heat waves. Dr. Marc

Verny wrote to Coquin about the deaths of two of his geriatric patients at the Salpêtrière Hospital in Paris's fifth arrondissement, both of whom died with fevers exceeding 106°F "with no clinical explanation." Verny concluded his brief report by noting: "In sum, the problems encountered, likely related to 'heat stroke,' are to be linked to specific climatic conditions. There is a risk of these cases multiplying, notably on wards with elderly patients or patients with neurological or psychiatric conditions that inhibit communication."[19] Later the same day, reports of similar deaths came into the DGS from the Hôpital Saint-Joseph in Paris's fourteenth arrondissement and the Hôpital Bichat in the eighteenth.[20] Coquin offered a terse response: "Yes, this conforms to the information we've received from the DDASS [regional health departments]. I think we need to set up an observation network by the beginning of next week."

By the close of business on Friday, Coquin had issued a press release to the regular media outlets, including print, radio, and television. The release indicated that "France is currently experiencing a heat wave that is capable of producing serious consequences for people's health," and offered a litany of suggestions for avoiding the heat's most extreme consequences. It identified newborns, young children, and the elderly as the populations at highest risk, and counseled concerned citizens to hydrate constantly, to avoid going out in the heat of the day, to wear light clothing and sun hats, and to avoid smoking, alcoholic beverages, and excessive physical effort. The release also listed the primary symptoms of heat stroke, advising anyone witnessing such symptoms in their relatives or friends to give victims something to drink and to force them to rest. "In the event that these symptoms persist," the release directed, "contact a physician."[21]

Coquin gave an interview to the local Paris paper *Le Parisien,* in which he expressed the concern that the country might experience hundreds of excess deaths linked to the heat.[22] But the national media failed to take any interest in the press release, and the InVS, the agency charged as a public health sentinel, reported for the weekend of 9 and 10 August that there was "nothing of note" happening nationwide.[23] Meanwhile, as national health authorities subsequently established, more than 6,500 people had already died from the heat's effects. Among them were seventy-two-year-old Gérard, who lived alone in a sixth-floor walk-up in the tenth arrondissement, whom first responders had taken to the Hôpital Salpêtrière, where he died on 4 August; Claude, a seventy-one-year-old who died in a full-care nursing home in the nineteenth; and André, a sometimes homeless sixty-nine-year-old who was then living in an attic apartment in the thirteenth.

Later investigations have suggested that the media ignored the press release because there was insufficient follow-up from the DGS. National news outlets receive hundreds of such releases in their in-boxes every day, journalists argued: the ones they take seriously are those that are accompanied by a telephone call from an agency authority.[24] Reporting on the heat throughout this period was prevalent but never urgent, despite mounting evidence of the heat's effects on the body. As early as 2 August, for example, one vacationer interviewed at a highway rest stop described the weather as "terribly hot," noting further: "I have been having an asthma attack, since noon it started. It's intolerable."[25] The media also recognized the ways in which the conjunction of the heat wave and the traditional August vacation made visible a series of human vulnerabilities. Stories about the homeless in Paris appeared on several television stations in early August, pointing to the difficulties of life on the streets at a time when the city's shelters and charitable organizations serving the homeless population were closed because of short staffing.[26] Some reporters even signaled an awareness of something menacing in the high temperatures. On 3 August, TF1's meteorologist Catherine Laborde—who had described the anticyclone as "protecting" France from bad weather two days earlier—now described the system as a "barrier," which was "preventing" cooler air and cloud cover from entering French territory. That same night, TF1 led with a story noting that the heat was becoming "difficult to take for the most fragile among us—that is, the elderly and children." By 5 August, TF1 aired an interview with Patrick Pelloux, an emergency room physician at Paris's Hôpital Saint-Antoine, who argued that "we'll certainly see consequences for mortality" as a result of the heat.[27] Several days later, firefighters reported a higher-than-usual number of calls related to heat stroke.[28] Yet these stories offered little in the way of concrete recommendations for coping with the heat, and moved on quickly to other stories that presented the weather with a smile and a shrug.

Viewed in hindsight, both press coverage and health surveillance during the heat wave's early stages present a fascinating combination of recognition and inaction that came to mark the fever pitch of the crisis in the following week. It is clear that the summer of 2003 presented the French with a period of extremes: suffocating heat waves, forest fires, landslides, avalanches, livestock die-offs, and a literally melting infrastructure signaled that the heat and drought had left no corner of the hexagon untouched. By the end of the first week of August, a human toll began to emerge with the deaths of a dozen elderly in Spain as well as a handful of deaths reported through official channels in France, indicating that the weather had begun to mark more than the natural landscape and the built environment. Reports touched on a range of potential vulnerabilities: short-staffed hospitals, an aging population, a government on vacation, shelters and parks closed. Health officials clearly recognized the emerging threat, Coquin even noting in one e-mail on 8 August that "the heat is beginning to present real problems."[29] On the same day, emergency room physician Philippe Hoang at the Hôpital Avicenne in Bobigny was alarmed to note four deaths from heat stroke, and notified *Le Parisien* that he had "never seen anything like this."[30]

And yet none of these clues translated into a meaningful general health warning at any level. Quite the opposite: as the Paris Fire Brigade's communications director later testified, on 8 August the Paris Police Prefecture, acting on the orders of then interior minister Nicolas Sarkozy, instructed the Fire Brigade—which typically handles first response in emergency medical cases, and was therefore on the front line of the crisis—"not to issue any alarming message and not to give the number of deaths" linked to the heat wave to the press.[31] The health warnings that did air on mainstream news outlets were directed at pet owners and parents of newborns. The elderly were mentioned several times as a population at risk in extreme heat, and featured prominently in a few stories. But overall the impression that such coverage leaves is that media outlets understood the heat wave as an inconvenience above all else. Yes, it brought some risk, but nothing that would prevent life—and the August vacation, for both the petit bourgeois in his camper and for the high-ranking minister—from proceeding unabated.

A Medico-politics of Disaster

All this changed on Sunday, 10 August, when TF1 broke the story of the heat wave as an emerging public health disaster. A day earlier, in the same story that featured the interview with Coquin, the daily newspaper *Le Parisien* had run an interview with Patrick Pelloux at the Hôpital Saint-Antoine, in which the physician stated flatly, "This is a crisis state." Pelloux had spoken with both television and print media in the previous days, but his concerns were largely ignored. Now, he sounded the alarm: "We've never seen this at the hospital, so many heat stroke cases linked to the weather. . . . Admissions are increasing by 20 percent each day. There are no more beds left." Yet even at this stage, TF1 was the only television station to pick up the story. In his interview with TF1, Pelloux told the film crew that "in four days, we've seen in the Paris region alone practically 50 deaths from the heat." He lashed out at political authorities, arguing that "at the level of the General Health Direction there's absolutely nothing happening—they even dare to say that these deaths are from natural causes. I completely disagree with this."[32]

This brief report—it ran approximately two minutes—transformed the story of the heat wave and introduced the principal terms of the debate that followed. It introduced audiences to Pelloux as the medical voice of the crisis, speaking from the trenches of the emergency room. It raised awareness of the heat's real (rather than potential) dangers for vulnerable populations by suggesting fifty deaths linked to the weather. And it raised, however subtly, perhaps the most enduring theme of the disaster: a polemical battle in which diverging sides assigned blame for the catastrophe and its management to their political opponents through recourse to contemporary political rhetoric. Pelloux imputed blame to the state for its inaction in the face of menacing danger. But the interview also established an acrimonious atmosphere between medical practitioners on the ground and state officials. By virtue of his position at the Saint-Antoine Hospital, Pelloux spoke with authority rooted in experience. Likewise, as the head of France's emergency physicians' union,[33] Pelloux was well connected with journalists, so he was able to command their attention. Yet Pelloux also had a history of employing hyperbolic rhetoric against the state.[34] An awareness of this tendency led some journalists, and certainly state officials, to regard Pelloux's claims as those of a self-appointed professional whistle-blower "who hadn't stopped crying wolf for years."[35]

Pelloux's testimony before the cameras—combined with an obviously rising death toll—forced the state to respond. And yet on the next evening's newscast, Health Minister Jean-François Mattei appeared to be utterly dismissive. When TF1's acting weeknight anchor, Thomas Hugues, asked whether the ministry had "underestimated the health consequences of the heat wave, above all for the elderly," Mattei responded: "I don't think so, not at all, that there has been an underestimation. Because to underestimate we would have had to have warning, but this heat wave was not something we could predict." He ended the interview by stating that the government would be establishing an information hotline the following day to respond to heat wave concerns, and he placed the onus on French citizens, rather than the state, to manage the weather's challenges: "Let's have a little more solidarity in this country. Let's pay attention to those around us."[36]

In 2003 the sixty-year-old Mattei had enjoyed a long history of success in his dual careers of medicine and politics. Trained in France's Centre Hospitalier Universitaire (CHU) system of academic medicine, he had served as a professor of pediatrics and genetics. A deeply committed Catholic, he also became an established authority in the emerging field of bioethics by marrying his religious sensibility to a rigorous understanding of the scientific issues at stake in a contemporary politics of life. He began his career as a politician with the right-leaning Union pour la Démocratie Française (UDF)

Party in 1989, when he was elected deputy for the Bouches-du-Rhône's second district. After President Jacques Chirac's landslide victory over the far-right Jean-Marie Le Pen in 2002, Mattei was named to head the Ministry of Health, Family, and Persons with Disabilities in the first government led by Prime Minister Jean-Pierre Raffarin. By most accounts, his public service as both deputy and minister was uncontroversial, and he enjoyed widespread respect as a medical and genetic authority in the revision of France's bioethics legislation in the 1990s, when the National Assembly took up the heady issues of cloning, genetic property, regenerative medicine, reproductive technology, and epidemiological screening in the wake of a range of scandals.[37] Even those who opposed his conservative politics lauded his character: one left-leaning colleague described him as a "perfect technician and impeccable humanitarian."[38] While he was far from an expert in epidemiology or disaster management, his medical authority, at least, was unimpeachable, given his strong credentials and long record of both publication and public engagement.

Mattei thus could likely have seized the reins of the emerging disaster if not for several critical missteps. His claims were both false and dismissive, if not flippant. To assert that the heat wave and its consequences for mortality were "not something we could predict" was misleading. The signs of a catastrophic drought and heat wave had been building in France for months, and by the first week of August the French meteorological service had signaled the extremity of the weather to come. Physicians in emergency rooms had been complaining about their working conditions since late July, and in interviews with journalists had noted the heat's effects on patients. Likewise, other branches of the government already had at least an inkling of the heat's effects on vulnerable populations. The Paris Fire Brigade had noted a significant increase in the number of heat-related calls it received in early August, as well as a number of deaths, and had communicated both to the Paris Police Prefecture, which suppressed the warnings. Even Mattei's own ministry had a strong—if delayed—suspicion that the heat wave would have significant health consequences, given Coquin's actions on the previous Friday and the interview he gave *Le Parisien* on Saturday.

Yet it was the interview's setting, Mattei's demeanor, and his concluding message that both set the tone for the state's response to the crisis and transformed the already contentious atmosphere into a full-blown polemic. Mattei appeared before the cameras not in a suit at the ministry or in a lab coat on a hospital ward; instead, he appeared at his vacation home in the Var in southern France, dressed in a black polo shirt (see fig. 3). Standing in a verdant clearing against the background of lush trees, with one of the few breezes in all of France blowing through his hair, he presented the air

FIGURE 3. Jean-François Mattei during an interview with TF1, 11 August 2003. Courtesy of Inathèque.

of having been interrupted on the golf course by a trifling problem. To his credit, he emphasized the importance of prevention in cases of heat stroke. But he repeatedly insisted that the state had the situation in hand and that, if anything, emergency rooms were better prepared for the crisis than they had been in years past: "The government is handling it, it's attentive, because each time we have extreme situations, a cold snap or a heat wave, the lives of the most vulnerable persons are in danger, and that's the reason why we have developed an appropriate plan."[39] But at the same time that he asserted that the state had the crisis in hand, he also began to shape the state's official message on the disaster: as a meteorological extreme it was beyond the state's control; it presented difficult circumstances for some but was merely a nuisance for most; and the real problem was a failure of community, rather than the government. The state could not hold the hand of every vulnerable subject—instead, community solidarity needed to pick up the slack.

The "polo" interview threw fuel on the heat wave's political fire. Mattei and his staff noted after the fact that the interview's staging was intentional, designed to relay an air of calm to an anxious public.[40] Juxtaposed with footage from hospitals of exhausted emergency room staff running from patient to patient, clips of patients arrayed in gurneys in hospital hallways because of a lack of space, and shots of first responders rushing victims into hospitals, the image of a casually dressed minister brushing off criticism framed the state's response as unserious. The image quickly became, as *Le Parisien* later put it, "the symbol of a tragic divide between the disaster of the heat wave . . . and its nonperception by the Minister of Health."[41] Worse, the interview with Mattei suggested not only a failure to recognize the disaster, but a dismissal of its potential consequences. A collapse in solidarity was the real culprit, for Mattei, because the deaths occurred on the fringes of society to those who were beyond state help. The physician Jean-Louis San Marco, who had experienced the devastating heat wave that struck Marseille in 1983, was vacationing in Italy at the time, but was moved to call the editor at *Le Parisien* with a grave warning: "We are faced with a human tragedy, a massacre undoubtedly never seen in France. And yet the impression that reigns is one of radio silence. This makes me want to scream! Right now elderly people are dying from the heat, but indifference is the order of the day because these are hidden, invisible deaths. Yet I assure you that these deaths are not natural, as some are saying, but for many are avoidable."[42] A French webzine, *Fluctuat*, went further in a biographical article, arguing that "Jean-François Mattei's political career ended with a Lacoste polo. . . . From the garden of his vacation villa, the health minister denied before millions of French that the heat wave of the summer of 2003 was a massacre of the elderly."[43] The national daily *Libération* considered the polo shirt to be a symptom of the ministry's mismanagement of the crisis: "Again Jean-François Mattei has missed the mark. The professor of medicine appeared unfazed when, in a polo, he wanted to be reassuring on TF1 on August 11." The article contrasted this nonchalance with Mattei's "frantic" demeanor several days later, then his eventual "opportunism" and "clumsiness" in accusing his own administration of mishandling the disaster.[44] Even years after the disaster, a number of my informants recalled the "polo" interview, shaking their heads in disgust as they addressed the minister's comportment during the crisis.

Mattei's casual appearance during the interview was deliberate, but the attempt to calm the public failed dramatically because the scene was set amid coverage of a crisis state as the nation's hospitals, morgues, and funeral parlors struggled in the face of a death toll that had already reached 8,000. Although it took demographers weeks to calculate these figures, it was clear at this point that a serious health disaster was under way. News broadcasts reported a three-hour wait at all of the capital's emergency rooms. Footage revealed physicians and nurses struggling to move through hallways packed with patients on gurneys because of overcrowding (fig. 4). Other stories showed firefighters alternating between carrying patients into hospitals in a flat-out sprint and hosing down hospitals' exterior walls in an attempt to cool them down several degrees (fig. 5). Meanwhile, news came into the DGS that the sheer number of bodies had begun to overwhelm funeral directors and that facilities for holding the dead were "saturated."[45] Emergency room

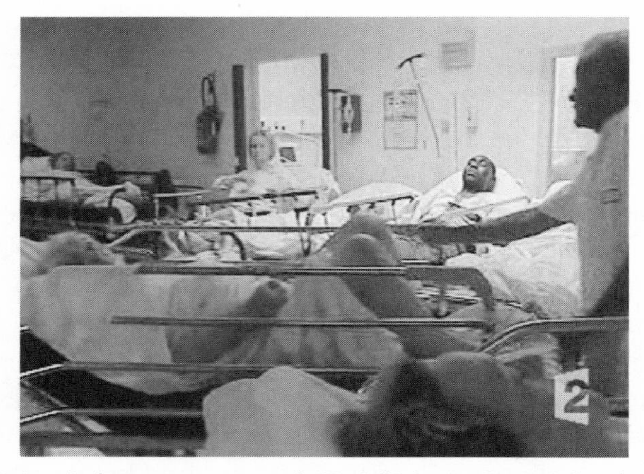

FIGURE 4. Patients in the emergency room at the Hôpital Saint-Antoine, Paris, broadcast on France 2, 14 August 2003. Courtesy of Inathèque.

FIGURE 5. Emergency vehicles amassed outside the Hôpital Lariboisière, Paris, broadcast on France 2, 12 August 2003. Courtesy of Inathèque.

and critical care personnel reported staggering increases in both admissions and mortality. Jean Carlet, an intensive care physician at the Hôpital Saint-Joseph, sent an e-mail to Lucien Abenhaïm, the Director-General of Health, late in the evening of 11 August in which he noted: "Emergency rooms, at least ours in any case, are completely inundated with patients, who are at times 4–5 days on a stretcher in the emergency rooms. We have opened emergency beds today in the hospital, and have completely shut down any programmed activity. I don't remember a situation like this in 25 years in the ICU. The situation is really quite serious."[46]

The "Massacre": A Cascade of Horrors

As Mattei returned to Paris on 12 August, the contrast between his rehearsed calm during the TF1 interview and the pandemonium of the hospitals established a major story line that haunted him through the catastrophe: that of a central administration ignoring those in close contact with the heat wave's victims. The hotline that Mattei had announced in his TF1 interview received over a thousand calls before it even officially went live at noon. Overnight reports of accumulating bodies in funeral services and morgues forced a realization that the disaster had reached a new level of lethality. One epidemiologist from the InVS reported that the Paris city morgue alone had taken in over a hundred bodies in a single day, which she described as "unheard of." The August vacations compounded the problem in two ways: a lack of personnel slowed the processing of the bodies, and families who were away were unable to claim their lost relatives' bodies.[47] That morning the Paris morgue had a hundred places remaining for new bodies. But in the course of the day, 185 new deaths were registered; on the next, another 217.

Officials began to characterize what they were experiencing as a *hécatombe*—a "massacre" or "slaughter." As the hospitals struggled to keep their patients alive and to disseminate information about prevention to the public, the Health Ministry and the press now became utterly preoccupied by death and its management. Officials recognized the looming threat of the approaching holiday weekend, which promised to slow the administration of death—from the clerical processing of death certificates to the physical burial of the dead—to a crawl. The normal legal requirements for transporting bodies into and out of the Paris morgue are quite stringent, requiring extensive police and medical documentation. Hygienic exigencies now forced a relaxation of these rules, minimizing the bureaucratic hurdles to the processing of bodies through the morgue. The Health Ministry went so far as to consider—before summarily rejecting—the possibility of cremating bodies that remained unclaimed after several days in order to free up space.[48]

Managing the living and the dead was by now taking its toll on health workers. Emergency room admissions had reached three times the normal rate in Paris hospitals, which were forced to call in the Red Cross and the French Army for assistance.[49] Footage on TF1 showed Red Cross volunteers dashing into apartment buildings and up staircases, sweat pouring from their bodies, to retrieve victims.[50] Paramedics and ambulance services in Paris reported mortality figures between one and a half to three times the normal rate for early August.[51] By 13 August, hospitals and pharmaceutical wholesalers in the Île-de-France reported major shortages in saline and other IV hydration fluids, while a shortage of hearses forced hospital authorities to seek authorization to use ambulances to transport the dead—a normally prohibited practice.[52] Some police units even ran out of latex gloves for handling bodies.[53] When Mattei returned to Paris and began touring hospitals, he was met with hostility by some health workers: one nurse at the Pitié-Salpêtrière Hospital rebuffed his greeting, shouting: "What a shame! No, I won't shake your hand. It's a real shame! There's no ice for cooling down the patients, there's nothing at all, and you come, what, fifteen days later?"[54]

But it was witnessing such pervasive death that proved most crushing. The coming days began to hint at the extent of the "massacre" that was taking place. In one interview, a journalist asked a Red Cross worker, "Is this the first time you've been faced with this many deaths?" The worker responded with a flat, exhausted expression: "This many? Yes. . . . It's a lot."[55] *Libération* profiled a Parisian physician who claimed that on 11 August he worked from eight in the morning until eleven at night, conducting twenty-five house calls and signing two death certificates. "And I'm just a neighborhood general practitioner. . . . We're running everywhere we go. The hospitals are swamped." As Paris hospital director Rose Marie van Lerberghe put it, "Our teams are used to confronting sickness and death, but not in these conditions. We have to recognize that at this point, for our staff—who, all of them, thought of a grandmother, of a parent—seeing people die was very rough. During the three hours that I was at Avicenne, four people died among the patients. It was very difficult to take."[56] Even those who managed death for a living were struggling in the face of this unprecedented mortality. In one funeral director's words, "It's atrocious. We can't organize a decent funeral, I've never seen this." He and his colleagues "don't even know where to look next," and "priests are running back-to-back funerals."[57] At the Paris morgue, one mortician described the scene as "madness," noting that he had seen one body that had been stepped on by morgue workers in the rush to process bodies: "It was very clear, someone had stepped on him. He still had the imprint of the shoe [on his face]. I did what I could to restore his face."[58]

After 12 August—a day on which, demographers later calculated, nearly 2,200 deaths occurred throughout France, and several hundred in Paris alone—the capital region had run out of places to store bodies and vehicles to transport them. For several days, the Health Ministry attempted to negotiate with municipalities that ran public ice rinks near the capital for storing bodies until they could be reclaimed. One official noted that "this solution seems to us to be ethically preferable to the use of refrigerated trucks that are normally used for transporting food."[59] This idea was rejected only when officials realized that such facilities were normally closed in August and would take too long to refreeze in such conditions. French law also prohibited the use of refrigerated food trucks for the transportation or storage of the dead.[60]

The public health danger posed by thousands of bodies soon trumped all other concerns. At this point Prime Minister Jean-Pierre Raffarin had declared the *plan blanc* or hospital emergency plan, opening all hospital beds in the country and calling physicians and nursing staff back from vacation. The temperature also broke that day. Maximum temperatures were still somewhat uncomfortable, but the life-threatening highs of the previous days had calmed to the 80s F (high 20s C), thereby stemming the tide of the sick and the dying. But the number of bodies produced during the heat wave proved overwhelming, forcing the suspension of ordinary regulations for managing death. Authorities erected refrigerated tents that maintained a safe temperature for hundreds of bodies. By 14 August, the Paris Police Prefecture noted that even working around the clock, "burials and cremations cannot keep up with the level of deaths," and along with other municipalities and departments, the city had extended the maximum delay between death and burial from six to fifteen days. But such a delay also entailed "a massive engorgement of sites for the accommodation of bodies." The city commandeered the food storage warehouse at Rungis on the southern outskirts of the city for the storage of more than 700 bodies at 5°C (40°F), and, forced by circumstances to suspend the law, leased or purchased refrigerated food trucks for further storage of hundreds of bodies in a parking lot in the southern suburb of Ivry: one truck, stripped of its decals, still bore the silhouette of the name of the butcher from whom the city purchased it. Yet within two days, the prefecture announced that even with these extraordinary measures, certain bodies posed a further problem, one that hints at the grisly facts of deaths from heat waves in particular.

> The major problem comes from the numerous bodies of people who lived alone, discovered at home in a deplorable state, notably in Paris where some 280 bodies in this condition since 6 August (and 90 on the single day of 15 August) have been documented. These bodies cannot be maintained at a temperature above freezing and necessitate the intervention of hygiene crews from the City of Paris who remove the body and complete a disinfection of the site. The Prefect of Police has decided this morning to rent 4 refrigerated trucks (capable of maintaining a freezing temperature), which will create, after an interior retrofitting, a capacity for accommodating at least 180 bodies.[61]

A recollection noted by the anthropologists Catherine Le Grand-Sébille and Anne Véga in a brief volume on the heat wave's effects on funeral workers and the victims' families portrays in vivid detail what the discovery of such bodies was like. They describe the interviewee as a "young woman of twenty, confronted here with the unthinkable." Upon returning home from vacation, the woman discovered that her upstairs neighbor had died during the heat wave and had remained there undiscovered for weeks. On her way home, another neighbor had called to tell her the news, but had no way of describing what awaited the woman in her own apartment.

> I screamed in horror at seeing on the ground a pool of dried blood, blood from a body, everything . . . urine, blood, everything. It had trickled down the walls, through the ceiling, I had paneling on the ceiling, so it passed through the slats. I had a kitchenette, sort of a bar, I had vases [on it] that were full of liquid, ugh. . . . When I got home, I had to vomit. I had lots of friends who had come, I grabbed a bag with some things and we closed the apartment. This lady, in fact, had died on the ground, directly on the floor [above]. In my apartment, I took my telephone to call the firemen to find out what had happened. And it was on the telephone that the firemen said to me, "Describe what's there." I was above the substances, and I said to them, "It's red, it's black, it's dried." "But of course, young lady, that's blood, that's urine, that's all the bodily fluids," the way the firemen speak, okay. . . . And then he said, "Well, yes, she drained, you see, she emptied out." And then he told me how they found her. . . . At this moment on the telephone, I was very, very sick. I was trembling, sweating, vomiting. . . . I couldn't do anything, I was nauseated, I spent practically the whole afternoon under the shower washing myself.
>
> In the apartment, the smell was still so strong, even months later, it never went away. It was impregnated in the couches, in the bed, everywhere, the smell was really . . . I'm retching. I threw up every time I went there, for several months, now I only retch.[62]

The city provided a disinfection service. She described the workers as "excellent," but also troubled by what they saw: although they wore gas masks, "they left retching." And despite cleaning the entire apartment with boiling water and bleach, the apartment was still uninhabitable. The cleaners had removed the big pools of fluid, but blood still stained the grout between the tiles, and the smell persisted. Her insurer described the situation as a "juridical void" and refused to pay.[63] Her only recourse was to sue the deceased woman's daughter for damages, a case that established new legal ground. Taking up one of the central mantras of the heat wave—that vacationers had ignored their vulnerable relatives and had therefore abandoned them to preventable deaths—her attorneys argued that had the daughter "been in daily contact with her mother concerning her health, she would have rapidly noted that she had died and the products of decomposition would not have deteriorated her apartment." In 2009, six years after the death, the court ordered the deceased woman's daughter to pay some 12,000 euros in damages.[64]

These discoveries were often gruesome. The combination of the intense heat and the vacation period meant that many returned from their travels to a horrific stench, and sometimes worse, in their buildings. Le Grand-Sébille and Véga recount the story of one funeral worker who was called to the scene by a hairdresser. Upon reopening her ground-floor salon after her vacation, she discovered that it was covered with maggots, which had spread from the body of a man who lived two stories above the shop.[65] News reports described entire apartment buildings that had to be evacuated because of the pervasive smell of death.[66] In my fieldwork studying the social histories of Paris's abandoned victims, I heard dozens of stories of these horrific discoveries: the concièrge in the thirteenth arrondissement who feared the worst when she approached one elderly resident's door and saw "a giant fly" emerging from a ventilation duct above his door; a neighbor in the twelfth who alerted authorities when she was assaulted by the smell of an elderly couple who had died in their apartment; the manager of a decrepit residential hotel in the eighteenth who checked on a tenant whose body was blocking the door when he tried to enter.

A 2003 documentary titled *La mort en face* (Facing Death) details a day in the life of those who collected bodies for Paris's municipal funeral services during the heat wave.[67] In a normal August, the services transport the bodies of the 38 people who die each day in Paris. When he realized that the services had transported 790 bodies between 11 and 13 August alone, François Michaud Nérard, the division's general director, called the television station France 2 to film his teams in action during this "catastrophe." His goal was to establish testimony of the exhausting, morally draining, and sickening work in which his workforce engaged both during the crisis and in periods of relative normality. But Michaud Nerard also argued that "we need to recognize that a death is not a statistical unit. It's a being of flesh and blood, which has social links, which has loved ones. . . . Behind the numbers, there is a person, a human reality."

Despite this emphasis, the film focuses less on those killed by the heat wave and more on those who are tasked with their management. The narrative begins early on the morning of Sunday, 17 August. A three-man crew has arrived in the Rue de Richelieu in Paris's second arrondissement to add a body to the ten that they have already collected that morning. A dreadful smell emanating from a tiny attic apartment has led the building's concierge to call the police in order to document a likely death. The workers—two of whom were hired by the city just a few days earlier—ascend a dark staircase. With a can of air freshener in his hand, the police representative opens the door to the squalid apartment, where the team discovers a body that has been decomposing for roughly a week. Wearing masks and gloves, the workers wrap the body in a bag, tag it with identifying information, and awkwardly lug it down the steps to the waiting truck. They then haul their decaying cargo to Ivry, where the bodies are loaded into awaiting refrigerated trucks whose makeshift shelves of cinder blocks and plywood allow for the storage of twenty-four bodies in each.

In another vignette, the cameras follow another exhausted team's processing of a body. The members of the crew have been working around the clock, and have not seen their families in days. As they ascend a staircase, the narrator notes, "each step takes them closer to an intolerable smell." They enter the apartment, but cannot find the body, following the smell "in some horrible game of hide and seek." Eventually they find it in a bedroom. They document and remove all objects of value from the body, then lift it from the bed. Again, the body has been decomposing in the heat for days: as they lift it, pooled fluid is revealed in the sheets and mattress. The team's leader, clearly experienced in these matters, explains that he would rather work a suicide in the metro than the heat wave's grisly scenes, "because then there's no smell." He also complains that he has had to throw away two pairs of shoes already in the course of the weekend.

Several years after the heat wave, a funeral official told me that the catastrophe completely overwhelmed his department. He had just departed on vacation on Friday, 8 August, and heard from staff over the weekend that mortality was ticking upward. Abandoning his trip, he returned to Paris at the same time that mortality spiked the following week. Bodies were piled everywhere atop one another. He was forced to call in reserve staff from other parts of the country that were not experiencing excess deaths—Brittany, for example, where temperatures were much cooler. But faced with such conditions many of the new arrivals did not last half the day. As a result the regular staff worked at least fourteen-hour days processing the bodies.

This funeral director emphasized that the smell of death caused the greatest suffering for the staff. Many of the bodies were not discovered until a week or more after their deaths when the smell began to pervade apartment buildings. He took great pains to explain how crushing it was to see his workers, with lit cigarettes in their mouths, carrying out bodies. Their smoking was not out of disrespect; instead, it was the only way that many of the funeral workers could diffuse the smell and avoid constant vomiting. He himself confided that for at least a year after the heat wave he could not eat olives, as for some inexplicable reason they were a horrible madeleine recalling for him the persistent smell of rotting flesh that marked the episode.

The film presents a jarring depiction of a largely hidden profession during a moment of extreme crisis. Utterly fatigued workers face what the narrator calls "unspeakable" conditions wrought by a combination of heat, isolation, and death behind closed doors. In one scene, morticians at the Paris crematorium at the Père-Lachaise cemetery must palpate the bodies for pacemakers, whose batteries could damage the crematorium by exploding during the process. The incident provides a startling insight into the nature of this work. Such examinations are normally completed by medical staff or are precluded by detailed histories provided by relatives. But in this case the workload was too great for medical examiners to tackle without help from nonspecialists, and most of the victims had been too socially isolated for the team to take accurate histories. And while surgeons might approach such tasks with more detachment, the job struck morticians and other funeral workers as an affront to their profession and its constant efforts to treat bodies with dignity.

The film takes an important approach to the heat wave in several ways. Although news media covered the heat wave and its death toll extensively, the film provides the only sustained look at one of the critical workforces involved in the management of the crisis. The film also humanizes those workers. Again and again, the viewer confronts enervated body collectors, morticians, and funeral directors engaged in a race against the clock as they attempt to process badly decaying bodies with some degree of dignity. Finally, the film reveals the care and respect with which those workers approached their macabre task, even when faced with a staggering workload in egregious conditions.

Left, Right, and the Placing of Blame

The rising death toll—which officials constantly revised from estimates in the low hundreds to more scientific tallies over 10,000—initiated a political crisis. Patrick Pelloux had begun stoking political fires with an interview on 11 August with the television station France 2, in which he added to the growing sense of an unbridgeable rift between the Health Ministry and emergency room personnel. Earlier that afternoon, Mattei's office issued a press release in which it declared that there was a "perceptible" increase in the number of hospital patients, "but there is no major increase of emergency services. The difficulties encountered are comparable to previous years."[68] When confronted with this message, Pelloux erupted at the cameras: "These people are on the ground, these are not people who are in their offices in the Avenue de

Ségur in an air-conditioned room, who don't know what's happening. The personnel are doing everything they can, and patients are dying. That's the reality."[69]

Given Mattei's blasé appearance during his TF1 interview, it was easy for the Left to portray him as out of touch with the realities of the crisis. But Pelloux's position as a union head made him a familiar adversary of the ministry. The Health Ministry thus accused hospital staff of exaggerating the crisis; emergency staff responded by accusing the state of ignorance and inaction. Very shortly thereafter these actors polarized into political camps of Right and Left, sounding familiar partisan drumbeats. From the perspective of the political Left, a government obsessed with downsizing the role of the state in private life was the clear culprit in the disaster: a failure to provide hospitals with the necessary staff and resources, especially during the traditional vacation period, was the root of the problem. From that of the Right, a culture of entitlement had fed the nation's worst egoistic tendencies while putting the most vulnerable French citizens at risk.

As the crisis mounted, Pelloux announced to France 2 that the biggest problems he and his staff faced were the lack of beds for the sick and "the exhaustion of the emergency personnel." Julien Dray, a spokesperson for the Socialist Party, argued that the Raffarin government "had not taken the necessary measures to confront this catastrophe."[70] Others on the left agreed: Yves Cochet of the Green Party blamed the government for missing the signs of an impending disaster: "There is a responsibility in [the government's] failures of anticipation and therefore, for a certain number of deaths, perhaps several thousand."[71] In the National Assembly, the Socialist minority leader Jean-Marc Ayrault argued that the Raffarin government had "underestimated the seriousness of the situation on the level of public health."[72] As France 2 noted on 17 August, "A lack of personnel, of means, but also of prediction: for the Left, the government is the primary guilty party for the crisis." And Socialist spokesperson Dray argued: "Why did it take so long to implement the *plan blanc*? And we need to know why the hospitals find themselves in such a difficult situation today. These are budgetary consequences that were voted on by the current majority."[73]

One notable exception was the former Socialist health minister Bernard Kouchner, the founder of Médecins sans Frontières, who asked on Europe 1, "What kind of society is this where we turn to the government when the weather is hot or cold?"[74] This was a good characterization of the response to the crisis on the right, where an emphasis on individual responsibility trumped the idea that the state should provide all solutions. For the Right, the problem lay in a decline in moral values and in the laziness of the French character. Mattei had set the stage for this response during his initial interview with TF1 when he implored his compatriots to "show a little more solidarity in this country. Pay attention to those around us." For a right-of-center government that had campaigned just a year before on a family values and security platform, the disaster was tailor-made for reinforcing this idea that moral decadence and an absence of responsibility were at fault. Media depictions of young and happy families cavorting on beaches in the Mediterranean juxtaposed with the abandoned elderly in hospitals and morgues suggested that the culture of the August vacation had placed the nation's elderly in jeopardy.

Some on the right focused less on a general social cynicism and more on specific policy issues. Christian Estrosi, a deputy from the Alpes-Maritimes and hard-line conservative, declared unambiguously that the thirty-five-hour workweek—a major Socialist victory of 2000 that had taken effect in hospitals in 2002—was at fault. "It's the 35 hours, without preparation, without means, without financing, that is today that cause of the disorganization of the hospital"; Estrosi recommended an immediate repeal.[75] The journalist Olivier Aubry, writing in *Le Parisien*, agreed: "If only there were just the summer vacations! Without replacement staff, the great vacation periods have traditionally been a high-risk period in the hospitals. But with the implementation of the 35-hour workweek last year, the summer disorganization has gone up a notch."[76] Aubry even cited Pelloux out of context, noting the physician's dismay at the lack of hospital resources.

Pelloux rejected this idea out of hand. He vehemently opposed the notion that the thirty-five-hour week had anything to do with the crisis, insisting repeatedly that he and his staff spent the entire period of the disaster with their patients, with any number of physicians returning early from their vacations to do so: "These caregivers gave without counting, without noting their hours. That's why they were so enraged when politicians of little faith and ambushing editorialists attributed the health consequences of the heat wave to the reduction of work time at the hospital, even though we were giving everything—we felt contemptuous, outraged."[77] Others took the attack on the thirty-five-hour week less personally, considering it another example of political opportunism on the right. For the journalist Vanessa Schneider at *Libération*, this was more of the same from Raffarin: "Under cover of empathy with the elderly, the government has renewed its favorite pastime: the war on 'free time'."[78] In the political atmosphere that produced the satirical volume *Bonjour paresse*, the economist Corinne Maier's scathing attack on a stereotypical French laziness, such attacks were to be expected.

Rather than quell this polemical discussion, President Jacques Chirac fanned its flames. After spending the entire period of the heat wave on vacation in Canada, he finally addressed the nation on 21 August. As returning vacationers continued to find the dead—although the peak of mortality had passed a week earlier, deaths remained above normal for the rest of August—a tanned Chirac offered a highly anticipated televised speech. Yet the speech not only failed to express any regret about his administration's handling of the crisis, but also harangued the French for abandoning their compatriots in their time of need:

> The tragic consequences of the heat wave show—and everyone among us must be conscious of this—the extent to which it is necessary that our society become more responsible, more attentive to others, to their problems, to their suffering, to their vulnerability. Many fragile people died alone in their homes. This tragedy sheds light once again on the solitude of many of our elderly or disabled citizens. Family solidarity, of course, the respect owed to elderly or disabled people, neighborly relations, and community action in this respect are indispensable for life in society. Individually or collectively, it is all together, with each in his place, that we should restore to these relationships and values their sense and their strength. Elderly and disabled people should be able to count on the solidarity of the French.[79]

This sentiment called on the same media representations that began this chapter: as the young and affluent headed south to the beach, they left the poor and elderly to suffer and die. As Philippe Douste-Blazy, director of Chirac's UMP Party, declared, "I would also say that [Chirac is] right to remind us of a bit of solidarity. We live more and more in an egoistic society that abandons its isolated elderly and I think that we all have to examine our consciences."[80]

But rather than blame a changing society, the Left insisted that the government still needed to take the lead. Annick Lepetit, speaking for the Socialist Party, responded to Chirac's speech by noting that "the president of the Republic has called on national solidarity. He's right, it's true that this is everyone's business, but I think that the state needs to be the first to provide an example."[81] Several days later, the left-leaning daily *Libération* issued its own scathing response:

> In Paris we have seen scarcely any government, and in the entire country, all these bodies of the old in morgues that are overflowing. We have heard these inept mumblings of Minister Mattei drawing up, in bold type and on legislative paper, this funeral pyre where the public's health will be burned tomorrow; there has been a sulking Raffarin pointing, to cover his incompetence, at these hypothetical "partisan polemics"; we have witnessed, to return fire, without a doubt, the little Copé and Estrosi, with their dehydrated inspiration, who will whine about how "it's the 35-hours' fault." There was Chirac, with

his tanned cynicism, barely resuscitated and already torn between words and deaths, but with enough guts to steer a cargo estimated at some ten thousand cadavers. Against or alongside these, but like an echo, there was Kouchner, all sobs and wavering voice, mounting a warhorse of good sentiments to launch an assault on a fatal national egoism and to vilify not the incompetence of those in power, but the indifference to the dying behind the "closed shutters" of Raffarin. Is there no one this summer in the Socialist Party who can establish an explicitly political relationship between the everyone-for-himself liberal and the end of civic morals? No.[82]

Accounting and Accountability: Science, Media, and the Production of Consensus

For both Right and Left, the attempt to place blame for the disaster on an adversary had far less to do with demanding accountability than it did with an effort to score political points. Those on the left—having suffered significant losses in both the presidential and the legislative elections of 2002—saw in the heat wave an opportunity to hang the crisis around the necks of the Chirac/Raffarin majority. On the right, an attack on a society of entitlements provided a deflection from the health and interior ministries' clear bungling of the case. As one conservative deputy put it during a parliamentary inquiry into the disaster, the heat wave had become a social rather than a health crisis, to the extent that the nation was less interested in finding solutions than in attributing blame: "Today, it seems as if in this commission, we want to hang a guilty party in a public square: the mayor of Paris who didn't get back soon enough, the health minister in his polo, maybe the interior minister, even, by default, other politicians."[83] To this end, the closest the administration came to admitting any fault was the forced resignation of DGS director Lucien Abenhaïm on 18 August, who immediately cast himself as a scapegoat for broader mismanagement of the crisis.[84]

In contrast to the media debate, official inquiries into the disaster in its aftermath promised to accomplish three things: to establish the facts of the disaster, including the precise number of deaths; to pinpoint what went wrong when; and to provide recommendations for the future. As Thierry Boudes and Hervé Laroche, two organizational management scholars in Paris, have argued, official reports on the disaster attempted to depoliticize the crisis in several ways. Their analysis of a number of reports, including those by health agencies, ad hoc committees, and parliamentary bodies, revealed a surprising degree of consensus, almost exclusively because of the nearly identical narrative strategies of the reports' authors. As Boudes and Laroche note, most of the reports lauded the heroic work of hospital staff and first responders, while largely ignoring the government's handling of the crisis. They also emphasized the failures of bureaucracy, attributing the mishandling of the disaster to a dysfunctional system rather than to specific actors. Finally, they recommended specific reforms in the system to avoid a recurrence of the problem rather than focusing on blame for the 2003 disaster.[85]

As Boudes and Laroche argue, each of these rhetorical tactics "depoliticizes" discussions of the heat wave. The reports single out specific groups for praise, but condemn bureaucracy and systems as flawed. The emphasis on reforms serves to reassure a disquieted population that such problems can be avoided in the future. These steps succeeded to a degree in moving past the left/right cycle of blame that marked the weeks immediately following the crisis, but they were perhaps more vituperative and less univocal than Boudes and Laroche acknowledge. Although the reports make little to no mention of the thirty-five-hour week or the "egoism" of French society, some reveal a strong political bias that emerges in other ways. Most important, their conclusions established policy directives and the representation of the disaster in a way that cemented particular ideas in the popular imagination.

Parliamentary inquiries were themselves of course political entities, involving extensive posturing during the questioning. Socialists grilled Mattei and other Health Ministry officials, while conservatives lobbed softball questions.[86] They reversed roles when figures such as Pelloux were on the stand. But more interesting are the ways in which the scientific reports that emerged from the heat wave reinforced social knowledge about the disaster that often emerged first in media circles.

The first of these reports to appear was commissioned by Mattei in late August, and was authored by a group of four physicians. The authors, led by Dr. Françoise Lalande, effectively shifted blame from the state to private actors. They specifically targeted private general practitioners, arguing that they had effectively abandoned their posts by leaving on vacation without sufficient patient coverage: "These absences seriously affected the operation of emergency services—which were already overwhelmed—and ambulance services." The report also singled out the media for failing to disseminate warnings about the heat wave contained in press releases that were widely distributed.[87] Through its use of the passive voice, the report soft-pedals its criticism of the Health Ministry. For example, instead of accusing the ministry of being out of touch or insensitive, the report's authors write: "In retracing the chronology of the disaster, it is fitting to keep in mind the fact that it is always easy, *a posteriori*, to foresee the unfolding of events. . . . One therefore notes a significant gap between the perception of the sanitary authorities and the reality of the crisis on the ground."[88]

While one might read the report as a fair defense of the ministry's inaction, the report shows several critical flaws. The attack on general practitioners is based solely on anecdotal data. There is no information about what proportion of general practitioners were away from their posts during the crisis, only the "testimony of numerous individuals" about "the difficulty encountered by patients and their families with medical emergencies and those seeking office visits, even for a regular consultation, or for the signing of death certificates during the week of 15 August."[89] Yet during a public health emergency, it is easy to imagine that a physician might be difficult to track down for an "ordinary consultation" or a bureaucratic procedure. The attack on the media is also shortsighted, as it considers the distribution of a press release the end of the agency's responsibility. Finally, the report emphasizes repeatedly that certain facts "could not have been known at the time," ignoring that information about the seriousness of the crisis was already circulating within the Health Ministry's inner circles.[90]

Other official reports worked in subtler ways to shape representations of the disaster in the popular imagination by assigning scientific authority to particular images of the disaster that already circulated widely through the media. When the demographic analysts Denis Hémon and Eric Jougla of Inserm, the Institut National de la Santé et de la Recherche Médicale (National Institute for Health and Medical Research), released the official state report documenting deaths during the heat wave, in late September, they confirmed what the images that had pervaded the airwaves since early August had already indicated. The heat wave was a crisis above all for the elderly, and in particular for the elderly who lived in cities. Of the 14,802 deaths that Hémon and Jougla attributed to the heat wave, 12,210—or 82.5 percent—were experienced among those who were seventy-five or older (see fig. 6). Among the elderly, women experienced 70 percent higher mortality than normal, with men experiencing an elevation of 40 percent. And while the heat wave elevated mortality throughout the country, urban regions experienced particularly high mortality, with the Paris region witnessing more than double the normal death rate. With only 2 percent of the nation's population, Paris contributed over 7 percent of the death toll. The report also indicated that mortality risk was highest among those in the "least affluent socio-economic categories."[91]

Hémon and Jougla's report is nothing if not disinterested. It is a concise assessment of the facts of the disaster, based on a careful demographic analy-

TABLEAU III.1 : Répartition des décès par âge et sexe pendant la période du 1er au 20 août

	Femmes				Hommes				Total			
	O	E	O/E	O-E	O	E	O/E	O-E	O	E	O/E	O-E
< 44 ans	**538**	**547**	**1,0**	**-9**	**1 310**	**1 159**	**1,1**	**151**	**1 848**	**1 706**	**1,1**	**142**
< 1an	72	76	0,9		105	95	1,1		177	171	1,0	
1-14 ans	45	41	1,1		59	58	1,0		104	99	1,0	
15-24 ans	60	66	0,9		208	191	1,1		268	257	1,0	
25-34 ans	91	101	0,9		275	270	1,0		366	371	1,0	
35-44 ans	270	262	1,0		663	545	1,2		933	807	1,2	
45-74 ans	**3 896**	**2 852**	**1,4**	**1 044**	**7 345**	**5 939**	**1,2**	**1 406**	**11 241**	**8 791**	**1,3**	**2 450**
45-54 ans	646	543	1,2		1 566	1 255	1,2		2 212	1 798	1,2	
55-64 ans	995	695	1,4		2 070	1 633	1,3		3 065	2 328	1,3	
65-74 ans	2 255	1 614	1,4		3 709	3 050	1,2		5 964	4 664	1,3	
≥ 75 ans	**18 018**	**9 543**	**1,9**	**8 475**	**10 514**	**6 779**	**1,6**	**3 735**	**28 532**	**16 322**	**1,7**	**12 210**
75-84 ans	6 414	3 417	1,9		6 169	3 919	1,6		12 583	7 336	1,7	
85-94 ans	8 878	4 924	1,8		3 748	2 564	1,5		12 626	7 488	1,7	
≥ 95 ans	2 726	1 202	2,3		597	296	2,0		3 323	1 498	2,2	
Total	**22 452**	**12 942**	**1,7**	**9 510**	**19 169**	**13 877**	**1,4**	**5 292**	**41 621**	**26 819**	**1,6**	**14 802**

FIGURE 6. Distribution of deaths according to age and sex during the heat wave. Source: Hémon and Jougla, *Surmortalité liée à la canicule d'août 2003—Rapport d'étape* (Paris: Inserm, 2003). Permission courtesy of Inserm CépiDc.

sis. (Chapter 5 below explores this and other demographic reports in greater detail and with an eye toward problematizing the nature of such analyses.) It contains no editorial comments, and its recommendations are for the establishment of early warning systems and more efficient mechanisms for collecting data. There is no attempt to produce knowledge about social categories, and yet its end result is the establishment of a portrait of aggregate risk during the crisis that is circumscribed by highly specific social parameters. The typical victim, according to Hémon and Jougla's analysis, was an elderly, poor, urban woman.

This portrait of risk became the central image of the heat wave: a poor elderly woman living in a tiny apartment in utter isolation despite being surrounded by an enormous urban population. It fed into media and state narratives about isolation and abandonment—the sequelae of life in a selfish society and a collapse of family values and social solidarity—as the root cause of the disaster. Images that circulated during the crisis primed the public to expect such conclusions. Footage of overwhelmed hospitals showed emergency room beds filled with elderly patients, especially women. But the report confirmed what most had suspected all along: that the heat wave killed those who were already near the end of their lives, and that it killed those among them who were the most isolated and the most desperately poor.

The critical point here is not that Hémon and Jougla's report was inaccurate—far from it.[92] Instead, the point is that the report's findings serve multiple interpretations of the disaster, and the media-driven interpretation of the disaster as consequence of social atomization in particular. What appeared to be an unambiguous demonstration that more than four-fifths of the heat wave's victims were very old reinforced a story that had coevolved in media and political circles throughout the heat wave. In an atmosphere of global climate change, the mixture of an aging and shrinking population with a culture of selfish entitlement, according to this logic, made for an inevitable if shameful disaster. This assemblage of available facts produced a not untruthful but still largely constructed popular knowledge about the disaster, knowledge that also mitigated the state's botched response in several ways. If the disaster was an inevitable "act of God," then the responsible agent was nature, rather than the state. Moreover, although the heat wave represented the worst natural disaster in contemporary French history, by killing those in the last stages of life it might be considered to have had what demographers call a "harvesting effect" rather than a seizing of life. According to this theory, the heat wave's principal victims were those who had only months or weeks to live in any event. While more careful and earlier state intervention might have saved them, it would only have forestalled inevitable and proximate deaths.[93]

The heat wave was an extraordinarily complex event. It combined an unprecedented extreme weather system with an aging society, but also with an exactly coinciding vacation period that placed many neighbors, family, medical staff, and the state away from the disaster's most affected zones. It involved difficult and fractured communications among a number of agencies and actors, with an important turning point in the disaster occurring over a weekend (9–10 August) that further complicated organization and increased response time. A decidedly abnormal health crisis—the outbreak of Legionnaires' disease in Montpellier—distracted health surveillance teams from a banal risk factor—high summer heat—at a critical juncture. The disaster exploited a near-perfect ecology for wreaking the maximum damage. It therefore resembles the sociologist Charles Perrow's model of a normal accident more closely than many other so-called natural disasters. Writing chiefly of technological disasters and human error, Perrow points to the ways in which "tight coupling" and "interactive complexity" generate important vulnerabilities in contemporary technical environments.[94] Taken on their own, the factors that shaped the heat wave disaster were relatively slight. High temperatures affect nearly all communities at one time or another, and only rarely with serious consequences. Vacation periods happen without incident every year in most affluent countries. Bureaucratic communication failures are a staple of the contemporary state, and aging is a critical demographic characteristic of most industrialized societies, in which shrinking family sizes and increased life expectancy have produced an older population. It was only in their interaction that these factors produced a devastating outcome.

Labeling the heat wave a "normal accident" that resulted from interactive complexity and system failure does not alleviate the responsibility of individual actors in the crisis. There were, as this chapter has demonstrated, any number of moments in which decisive action might have made a difference, with information readily available to key stakeholders. Yves Coquin could have followed up his press releases with phone calls to journalists; the press could have taken the story more seriously when Pelloux and other emergency room personnel signaled a crisis state; Sarkozy could have released mortality data from first responders; Mattei could have declared a state of health emergency; Raffarin could have implemented the hospital emergency plan sooner. But the normal accident framework highlights the complexity of the disaster's making and its outcome. Where some factors were directly related—the Legionnaires' outbreak drew attention away from the health risks of extreme heat—others were purely coincidental, such as the arrival of the heat during a vacation period.

Yet media narratives, political rhetoric, and epidemiological data production reduce this complexity. For different reasons, each of these categories of storytelling tends toward simplification. The news media strive to produce a clear and compelling story line: abandonment of the elderly puts them at extreme risk. Such stories make us reflect on our own position: Who among us could not pick up the phone and check on an elderly neighbor or relative a bit more frequently? Offer to pick up a few groceries, or even stop to chat? Political story lines also strive for simplicity: a shortened workweek leads to understaffing, disorganization, and panic at the hospitals, while our decaying values lead to an overreliance on the state instead of community and family; or on the other side of the spectrum, a state shirking its responsibilities to its citizens has fractured the social contract and condemned its most vulnerable members to death. For epidemiological work, the tendency toward simplification has different motives but a similar outcome. Where is the aggregate risk highest? What is the most vulnerable population? In the case of the heat wave, the prominent risk group is undeniably the elderly, for physiological as well as social reasons. Aging bodies are increasingly vulnerable to heat in a number of ways. Declining sensory perception delays critical messages of dehydration or overheating from reaching the brain until it is too late. Increasing debility that accompanies aging increases social isolation, as does the outliving of a social and family circle. Poverty also often accompanies aging, exacerbating the risk. In a public health environment of limited funding, a policy focus on the group at highest risk—especially one that so clearly stands out from the general population—makes good sense.

Yet there are several problems with simplifying the causes of the crisis and the general portrait of the disaster's victims. One factor is the remainder of the victims. As chapter 5 below shows, a risk profile that emphasizes the vulnerability of the elderly above all else ignores the nearly 3,000 victims who did not fit this profile: nearly double the death toll from Hurricane Katrina, and nearly the same as the victim count of the September 11 attacks in the United States. It is an oversimplification that serves important purposes, and reduces risk for the elderly in future heat waves. Yet it also potentially exacerbates risk for other groups. Likewise, not all elderly are equally vulnerable to extreme heat. But this risk profile effectively makes everyone over seventy-five into a potential victim while virtually ignoring other risk groups.

Another important problem with the establishment of this aggregate profile of risk connects to the rhetorical frame that emerged from the state when first confronted with the disaster. It portrays those vulnerable to heat waves in a manner that precludes easy identification with them. A constant barrage of images in the media about the heat wave and its toll showed elderly person after elderly person in rows of hospital beds. Naked, frail, helpless, mouths agape, many appear demented and near death. It is a representation of aging, isolation, and poverty that is difficult to assimilate to ordinary experience. Although this is an important truth of the disaster, it amounts—along with the Raffarin government's emphasis on individual rather than state responsibility for preserving well-being—to a portrayal of the victims' marginality that emphasizes their physical, cognitive, and social segregation from the larger community. It is in some ways a final assault on their dignity: with the victims denied a "good death" by the fact of social isolation, such a depiction reinforces their distance by fixing them as a type alien to general experience, rather than as individual victims, and citizens, in their own right.

In the waning days of the heat wave, the emergence of another compelling tragedy underscored the ways in which storytelling framed victimhood and shaped a particular mythology of the disaster. A steady influx of bodies during the heat wave was one factor that overwhelmed public health and funeral services throughout the country, but especially in Paris. But another source of the problem was what the media quickly labeled the "abandonment" of bodies: the failure of relatives or friends to pick up the remains of the dead. These were the abandoned or "forgotten" victims of the heat wave, those with few social contacts who lived and died on their own, and who provided stark, if anecdotal, evidence of an atomized society.

As the rest of this book demonstrates, the media depiction of the heat wave's forgotten victims similarly reduces the disaster's complexity in disturbing ways. As journalists and documentarians have structured it, the story is emotionally wrenching: it is a portrait of a wealthy society that has abandoned its undesirables to abhorrent desperation. Yet a closer investigation of the many stories of the forgotten reveals complexities that the dominant narrative of social abandonment conceals, and sheds critical light on the long-term historical development of a series of vulnerabilities in French urban society that structured risk during the heat wave. It also, as the following chapters show, exposes the ways in which that dominant narrative served important rhetorical purposes for a grieving French public in the wake of unspeakable disaster.

2

Anecdotal Life: Isolation, Vulnerability, and Social Marginalization

It's a lot to talk about, Marie France, because I moved in six years after she did, so I knew her a long time. But by June 2003 I had no more relations with this lady, because when my husband had some work that involved a lot of travel, and we left for Malaysia for a year, she didn't like this. She didn't like this, she said we had done it intentionally, to leave her alone and all that. But no, it was for work. Work is work. So we had my daughter live in the apartment with her husband, they're very quiet, but she found problems. They made too much noise, they showered too late at night or too early in the morning. . . . So when we returned from Malaysia, she said "bonjour," "bonsoir," period. She was intelligent enough to know that she had done things during my absence that didn't please me. And I preferred not to argue with her. So our relationship didn't really exist anymore, we could say.

—INFORMANT ON MARIE, d. 12 August, twelfth arrondissement

Marie France was a fixture in the building where she had lived for as long as anyone could remember. She lived in a tiny apartment on the seventh floor of the building, with windows that looked directly out onto the Bastille Opera. The apartment was a *chambre de bonne* of about 100 square feet with a toilet and a rudimentary corner shower. Despite such basic accommodations, the eighty-eight-year-old took great care with her appearance. She regularly visited a local salon to maintain her dyed blonde hair, and she was always sure to wear full makeup in public. She was quite reserved according to her neighbors, polite but standoffish. Only after knowing one of her neighbors for years did she open up to her, sharing details about her family history.[1] She was quite proud of an older brother who had died honorably during the First World War. She had shown his photograph to the neighbor several times, along with those of her parents. She had been married once—perhaps widowed, although details were unclear on this—and also mentioned a life she had led in Algeria during the mid-twentieth century, which included an engagement to a soldier that for one reason or another failed to end in a marriage. In short, she lived a modest but independent life, at least until she broke her hip in June 2003.

At that point she stopped leaving her apartment regularly, going out only on occasion. Yet she still insisted on her self-sufficiency, rejecting all offers of aid from neighbors, who tried repeatedly to register her with the city for a program that provided free domestic assistance with cooking, cleaning, and shopping. The paperwork was too much trouble, she said. When neighbors offered to complete it for her, she dismissed that idea as well: she was perfectly able to take care of herself, and she resented the intrusion into her private life.[2] She was also cantankerous, as the above excerpt from my field notes indicates. Her neighbor told me that she had always kept her apartment in immaculate condition, but had recently let it go. She threw and broke things in her apartment in fits of rage. She became increasingly argumentative with her neighbors, pushing them out of her life.

Marie France's name figured prominently in an unprecedented article that shocked the readers of *Le Parisien* and its national counterpart *Aujourd'hui en France* on 2 September 2003.[3] The article included a disturbing list: a compilation of sixty-six names that the city of Paris had shared with the newspaper. It was a list of those who had died during the heat wave and whose bodies remained unclaimed by any relative or neighbor. While death is normally a private matter in France, the publication of these names in such a prominent forum had two principal motives. First was the possibility that such a list, which included the victim's full name and, where possible, the date of birth, could help authorities to locate the deceased's next of kin and allow for a proper burial. But close behind was the possibility of shaming family members who knew about the death but had failed to step forward to collect their relatives' bodies. In a moment when public officials were overwhelmed by death such a move seemed the only option.

For several weeks stories of unclaimed bodies had resonated in the press, initiating a media polemic. The press on both left and right christened them the "forgotten" or the "abandoned" of the heat wave, and considered them a repository of national shame in the wake of the disaster. The forgotten were a symbol of inhumanity: for the media and for many politicians, they were proof positive that the real cause of the heat wave's staggering death toll was not state negligence but an egoistic and uncaring society. A list of headlines gives some sense of the shame that surrounded the deaths: "450 Forgotten Deaths: We Are All Guilty" and "We Should Be Ashamed" (*Le Parisien*); "Forgotten Deaths and a Repentant Nation" (*Libération*); "French Barbarism" and "When Selfishness and Indifference Kill" (*Le Figaro*).[4] Such stories diverted the rage that many had expressed over the state's handling of the crisis toward apparently uncaring families and neighbors. *Le Parisien*'s list of names took things a step further, however: it deliberately identified the dead toward the utilitarian end of emptying the morgues and alleviating public responsibilities, on the one hand, and that of redirecting blame on the other.

A cartoon that accompanied the article shows a man and child walking past an imaginary statue memorializing the "abandoned"; as the child says, "Hey, Grandpa's name is there!," his father averts his eyes and replies, "Shhhh!"

The list and those who populated it became central to the story of the heat wave. They became a multimedia sensation in France, with widespread representation in television, print journalism, documentary film, and the plastic arts. The story of these deaths in anonymity underscored the pathos that the shock of so many deaths had introduced weeks earlier. The notion that so many deaths could occur in such utter isolation struck many French as a horrifying reflection on a culture of egoism that the heat wave had cast in high relief. But even more than the portraits of risk that emerged from epidemiological reporting (discussed in chapter 1), the story of the heat wave's forgotten victims reinforced the impression of their marginality. Tales of such loneliness evoked pity. But they also cast the victims as in many ways responsible for their own alienation, or even as the manufacturers of their invisibility.

This chapter explores this phenomenon in several ways. Beginning with the public ceremony that marked the burial of the forgotten, it explores the emergence of dual narratives of pity and shame that surrounded the event, which became a symbol of the nation's mismanagement of the disaster as a whole. Research into the life histories of the heat wave's forgotten victims reveals important tensions in these representations, which I argue came to constitute a principal social memory of the disaster. Stories in the press and those that various informants revealed to me in the course of my fieldwork yielded a phenomenon that I call "anecdotal life," in which a series of small notations of apparently random biographical moments defines the contours of a life history of these victims, one that leads inevitably to the victims' deaths in isolation. Such a strategy serves as a mechanism for fielding blame both for the state and for those who surrounded the forgotten, and with important implications for the production of marginality that continues to haunt these victims even in their deaths.

Thiais: Pomp and Circumstance in a Technical Necropolis

I brought him something to eat, and he never responded on the interphone. I found that strange. After that I went up to the apartment, because since he didn't answer I didn't know if he was sick or what, and I saw, you see above the door there is this vent that the wiring passes through? I saw a giant fly come through, and that made an impression on me. After I went down the stairs, I left his food on the landing, and I called the firemen. They went in through the window and found him next to his bed. He was dead, and his body had drained. It was 115 degrees in the apartment. And he was in a sweater. And really, when I saw him he was always in a sweater, and when I told him to drink, and I did some shopping for him . . . he would say that it's thanks to her that I can live another day or another year. Because his brother had died, and he let himself die.

INFORMANT ON ROGER, d. 8 August, thirteenth arrondissement

The list of forgotten victims established new journalistic ground, but it also introduced the major narrative frames that have shaped the history and memory of the heat wave. The funeral ceremony for those victims, held at the Parisian public cemetery in Thiais on 3 September 2003, reinforced those frames. Both the list and the ceremony grouped these deaths and invested them with new meaning: rather than a few dozen individual deaths, they became the phenomenon of the forgotten, henceforth making sense only as an aggregated list. Although the deaths of these individuals might have constituted tragedies in their own right, their grouping opened a new social category of death that underscored some of the central revelations of the heat wave and gave them a new importance. Among the most important imaginative frames the forgotten called to the fore were isolation, which played a key role in their deaths; the closely related theme of abandonment by state and society; and a search for explanation: how could this have happened in a country that had been founded on an ideal of fraternity, and imagined itself as a birthplace of human rights? Yet while the soul-searching that their deaths prompted was critical in drawing attention to the problem of social isolation in France, especially among the elderly poor, the explanatory narratives that emerged from the crisis effectively reinforced the invisibility and the marginality of the forgotten.

The bodies of the forgotten came into public consciousness through the work of François Michaud Nérard and an indefatigable team of funeral workers who coped with the mountains of bodies that the heat wave produced. As the director of Paris's funeral services department, Michaud Nérard noted in his memoir of the catastrophe that as the crisis in the hospitals began to wane it only intensified for his workers. Since 14 August, he wrote, "ALL the bodies that we had to pick up were bodies that had been found several days after death. They were people who were alone, fallen, struck down in their tiny apartments, in miserable *chambres de bonne* on the fifth or sixth floor. It was often over 100 degrees, the bodies in an advanced state of decomposition, full of insects." More than 300 of the bodies they collected were in this "horrible condition." The work was "physically but also morally exhausting," because for a profession "associated with slowness, respect for the body, attention to relatives and dignity, it meant going against nature and it was intolerable."[5] Normally an invisible labor force, France's funeral and cemetery workers struggled both during the crisis and in its aftermath. Transporting, identifying, and storing bodies; finding families and organizing funerals where possible; digging graves; and managing the tangled bureaucracy of death in unprecedented numbers were overwhelming for funeral services throughout France, but most particularly in Paris.

The disaster tested a system that has been in place in France since the Revolution. The secularization of the state in the late eighteenth century precipitated the charging of French municipalities with the management of death, and the burial of the poor in particular.[6] Until recently, communities buried these bodies in mass graves, a process that French law still allows, but one that municipalities no longer typically practice. The Thiais cemetery traces its origins to this requirement; it is, in the words of one visitor, a "cemetery of relegation," with few of the celebrity monuments that one finds at Père-Lachaise or Montparnasse.[7] Among its nearly 150,000 monuments, it counts over 3,600 individual tombs that serve as temporary resting places for those who have died in Paris without resources. In 1992 then-mayor Jacques Chirac authorized the installation of these tombs as a hygienic alternative to burial in the ground. They are a marvel of thanatotic technology. The poured-concrete sepulchers emphasize utility, reusability, and hygiene. They are identical, laid in rows, each with a slot on the top that holds an engraved brass tag bearing the occupant's name, birth date, and death date, where available. The tombs each accommodate a wooden coffin. Cemetery staff guide a hand-operated crane mounted on wheels to the tombs in order to lift and replace the heavy concrete lids (fig. 7). Each tomb is connected to a drainage system that evacuates any fluids that seep through the coffin, while an underground ventilation system sweeps away any noxious odors of decomposition.

In a given month, the cemetery receives about thirty bodies.[8] But in August 2003, over a hundred came into Thiais.[9] When I spoke with cemetery workers who had experienced the crisis, they repeatedly emphasized the exceptional nature of the period. Despite the efficiency of the sepulchers, the volume of bodies overwhelmed these workers. Many personnel returned early from vacation either through orders or of their own volition. Traditional in-ground burials and the transfer of bodies consumed most of the staff's time. In the words of one cemetery worker, they worked from early in the morning until late at night in the heat, with "not a moment's rest."

According to many assessments the staff at Thiais generally succeeded in handling a difficult situation. Conditions in the *secteur d'indigents* are respectful toward the dead, if antiseptic and mechanized. A comparison to the mass burial of Chicago's "abandoned" or "unclaimed" victims in the aftermath of the 1995 heat wave is telling. As Eric Klinenberg points out in his scathing assessment of Chicago's management of the crisis, the city buried its

FIGURE 7. Thanatotic technology: a mechanical crane designed to open and close tombs at Thiais to facilitate burial and removal, allowing for easy reuse. Photo by the author.

dead in a mass grave on the outskirts of the city. Workers excavated a trench 160 feet long and 10 feet wide, and dropped the forty-one remaining bodies of heat victims, each in a plywood casket, into the grave at a total cost of less than 100 dollars per body. A minister said a few words, then workers closed the trench. The city commissioned no tombstone or other marker to memorialize the bodies; a year later a private group commissioned a small stone monument. Witnesses described the event as reminding them of 1940s newsreels depicting European atrocities.[10]

By contrast, the ceremony at Thiais featured the head of state and other high-ranking officials. A long procession marked the funeral service at Thiais on 3 September. Many of those who arrived intending to commemorate the deaths of the forgotten were turned away for security purposes, as President Chirac and Paris mayor Bertrand Delanoë attended, along with Health Minister Mattei. Monsignor Patrick Chauvet of the Paris diocese and Dalil Boubakeur, rector of the Paris Mosque, represented two of the city's spiritual communities. One story described "a cemetery in a state of siege," noting that Thiais had "never . . . witnessed such a crowd," with official vehicles causing traffic jams and security teams blanketing the grounds.[11] As part of the ceremony, a mayoral aide recited the lyrics of the 1970 song "Quand ceux qui vont" of the chanteuse Barbara, and Michaud Nérard read the names of the victims aloud. In a twenty-five-minute ceremony, the group attended the burial of the first two victims, the forty-two-year-old Valérie Dumans and the eighty-year-old Paulette Moreau. The burials of fifty-five others followed in the course of the afternoon after the officials' departure.[12]

Media coverage of the ceremony was widespread. On TF1, chief anchor Patrick Poivre d'Arvor (who had spent the entire crisis on vacation) led with the story. "There was an homage to the unknown soldier of the First World War," he began. "There are now the unknown dead of the summer." The story focused less on the ceremony and more on the "bitter reactions" of those in attendance. Some forty emotional visitors registered their grief and anger at the very idea of the forgotten victims. One elderly woman told TF1, "It makes me sick to see something like this in France." Another told the network: "That one could abandon their parents, their mothers, their friends, or even their neighbors, it's something that's unthinkable. Our obligation is to wake up our consciences and to say, 'Stop'."[13] France 2 featured similar responses, in particular one young woman who expressed her shock at the extremes of an "individualistic society." Particularly moving were the comments of one man in his sixties: "I'm here as an individual; because I live alone, I know what solitude is like. I'm getting to know it more and more because I'm retired, and when one is retired, one is abandoned by everybody."[14]

The left-leaning daily *Libération* described the event as "a grand ceremony for those who died from indifference," quoting visitors who described the situation as "unthinkable" and "a total catastrophe," but focused on the political scandal of the heat wave.[15] *Le Parisien*, meanwhile, which had covered the heat wave crisis from a number of angles and had published the list of the forgotten a day before the interments, offered more extensive coverage. The interviews that the journalist Marc Payet conducted with those in attendance were univocal. Mostly seniors, the interviewees condemned what they considered to be a horrifying and dehumanizing series of deaths in isolation. "I don't want them to be abandoned in their solitude, like dogs, so I decided to come," claimed one seventy-three-year-old woman. Another, seventy-one, said that she "came to pray for them and their families, and to attest to my solidarity in this period of suffering." Another, sixty-seven, described the situation as "improper." She was "shocked," and "came to replace their absent children."[16] One sixty-three-year-old man expressed a sentiment that was emblematic of the national conversation about the forgotten: "I had to be here, because I think there's nothing more terrible than ending up alone, without anyone at their funeral. I have a sense of compassion for these people. I feel demoralized, overcome, by the number of families that have abandoned their elders. The only positive side of this tragedy is that we're finally talking about the solitude of the elderly. . . . These people finally have a right to respect, that's what's important."[17]

The burial at Thiais ended the public health crisis that the bodies of the forgotten posed, with the state simultaneously absolving itself of its mismanagement of the disaster. But as with the staggering mortality of the heat wave in general, the problem of death in anonymity presented France with a moral crisis that proved more difficult to resolve. News coverage of the funeral ceremony provided a glimpse into the way in which the heat wave cast the deficiencies of contemporary French humanism in high relief. Interviews with attendees delivered the main thrust of this story, which amounted simultaneously to a display of pity and finger-pointing. Yet the interviews also fell within the allowable narrative boundaries that had characterized the political debate over the heat wave since mid-August. Some issued attacks on an "egoistic" and "individualistic" society and blamed the predicament of the forgotten and other heat wave deaths on a collapse of social solidarity and an epidemic of selfishness that critics insisted increasingly characterized French society, while others charged the state with a catastrophic failure to protect its citizens. Both of these responses constituted reactions to a mass phenomenon; neither considered the conditions or circumstances of the forgotten victims as individuals—something that followed in the ensuing weeks and months.

Anonymous Death and a Crisis in Humanism: The Making of "Les Oubliés"

They were a very sad case actually. I was always very nice to him. I never saw her actually, because she was pretty much of an invalid. She couldn't hear, they had this loud TV. They heated with coal, they were extremely poor. . . . Monsieur C., he was so weird, everybody was wrong, everybody was bad, everybody was against him, whether it was left wing, right wing, young people, old people, everybody had something wrong with them. He had an enormous chip on his shoulder. Anyway, it was really sad.

INFORMANT ON CLAUDE AND MARCELINE, found 17 August (d. several days earlier), twelfth arrondissement

In the aftermath of the burial, story after story appeared in a variety of media that attempted to explore the problem of the forgotten as a window into the heat wave as a social catastrophe. According to this narrative, the heat wave was less of a scandal in its own right than it was a means of revealing scandalous conditions at the heart of French culture. Yet there are several important elements that characterize the public treatment of the forgotten, which are linked to ways in which this treatment attempted to homogenize a diverse group whose principal connection was death in isolation. The forgotten shared some important tendencies. They were generally older and poorer than average French citizens. They mostly lived in substandard housing and in marginal conditions. But the most important situation that they shared was their death in isolation. It was their condition as "unclaimed" or "abandoned" in death that structured—and continues to structure—representations of their lives. A public imagination of their condition began with their deaths and retroactively pointed any subsequent narrative of their lives toward the inevitable outcome of abandonment.

The forgotten constitute an anomalous group who represent an ordinary phenomenon. Deaths in social isolation happen regularly in France—on average, one per day in the capital. My conversations with Cécile Rocca, the director of the NGO Morts de la Rue (The Dead in the Street), highlighted the regularity of such deaths among France's most marginal citizens. The organization, founded in 2002, seeks to promote awareness of the violence of everyday life among France's homeless and those who live on the precipice of homelessness. Throughout the country in any given year, roughly one homeless person dies per day. The causes of death are wide ranging. Some die from the cold; others die from the heat. Some die from tuberculosis or other infectious diseases. Some die from their addictions. When I was in the organization's offices for one interview, Rocca received a call notifying her of the death of a homeless man she knew who lived in front of a grocery store in the third arrondissement. He was young—in his thirties—and had recently stopped drinking. The abrupt withdrawal from alcohol took his life. But most, Rocca assured me, died violent deaths: accidents, murders, and suicides marked the everyday life and death of the population.

But if death in isolation is an ordinary—and therefore invisible—problem, it was the concentration of such deaths that gave the forgotten their significance in August 2003. These deaths assume meaning as markers of social fragmentation, the collapse of solidarity, and atomization only when one considers them as a group. Yet efforts to shed light on the group in the aftermath of the disaster focused on members of the group as individuals. This effort to individuate the group resulted in a fractured narrative in accounts of the disaster and the idea of isolated death. Most of these stories sought to humanize the forgotten by illustrating their particular social histories as individuals. But there is an important paradox in the individuation of a collective problem. While some of these stories succeeded in developing sympathetic portraits of the forgotten as victims of both a disaster and an alienating society, many achieved the opposite effect. By developing a life story from the starting point of a horrific death, they cast these figures as the primary agents of their isolation. The effect of many such accounts was effectively a redistribution of blame for the disaster through a sort of marginalizing memory.

Where initial debates during the heat wave positioned either the state or an uncaring society as the disaster's major culprits, the subsequent narratives of individual subjects as abandoned, unclaimed, or forgotten tended to focus on these figures' instability, their moral failings, or their marginality itself as factors that drove their isolation and structured their vulnerability.

Some accounts reiterated the emotions that had surrounded the story of the forgotten as it had emerged: that their deaths were clear evidence of a social decline. The Catholic newspaper *La Croix* led this charge. In a story the paper published on the day of the Thiais ceremony, Dominique Quinio wrote that the "number" of the forgotten "shocks our conscience." The journalist signaled the city and its forms of sociability as a deadly environment for the elderly and the marginal: "The big city exacerbates solitude. . . . Modern life exalts the individual, at the risk of weakening family and social solidarity."[18] Letters to the editor of the paper echoed this sensibility. One woman in her eighties wrote that she had had only two contacts with others during the month of August: "In a month, that's not much." She noted that she had no close family, but that she had adopted "a small dog, also abandoned in the month of August." She signed the letter, "I hope you'll excuse this bittersweet letter (more bitter than sweet)."[19] Other articles adopted a scathing tone. One singled out for blame the "silent families who let their elders depart in incomprehension or in indifference,"citing a genealogist who noted that "there was a time that we called on genealogists to search for very distant legatees. Today, families are torn and reconstituted, so one calls on us to find even children or parents!" The story accused some of refusing to claim their relatives' bodies for "fears of inheriting their deaths and of all the administrative paperwork." These families simply "didn't want to hear any more about their dead."[20]

Other investigations revealed widespread disbelief at the problem of the forgotten. *La Croix* ran a story about All Saints' Day in the cemetery, which focused on the "'famous' solitary victims of the Parisian summer. Famous only because they had been forgotten. By their families, for many by an entire society." The author cited two visitors to the Thiais cemetery, who claimed: "We can't believe this. . . . It's truly unimaginable that so many people could be discarded."[21] A documentary on France 3 titled *Vieillir ensemble?* (Growing Old Together?) explored the same phenomenon through a focus on one victim, Adèle Angèle. Adèle was not buried at Thiais, nor did she appear on *Le Parisien*'s list, but this was more a function of geography than of her social condition: she had lived in the Paris *banlieue,* outside of the city's catchment. Otherwise, she conformed to the model of the abandoned victim. At age ninety-two, she had lived alone in her apartment since the 1960s. Adèle had never married and had no family. On 15 August, the fire department discovered her collapsed on the floor, a chair overturned, a half-eaten meal and an emergency telephone number on the table before her. No one knows exactly when she died, although it was clearly several days before she was discovered: the reporting officer noted that the heat certainly "doesn't help preserve the body." But despite her isolation, neighbors expressed shock at her death. She had always seemed so independent, they claimed, never wanting to impose on anyone. One neighbor expressed her horror at Adèle's plight: "I reject the idea that one can be alone at the moment of one's death, that's very, very sad. . . . The need to touch a human . . . for me, the idea of dying alone is intolerable." Another expressed her culpability in a matter-of-fact tone:

> I tell myself that I should have knocked on her door, that's for sure. But I didn't. . . . Quite simply, because I had absolutely never thought of doing so. It's unbelievable, but that's how it is. Quite simply, because this person was nearly invisible. You see, I don't know how to explain it, I have work, my husband and I have other things to do, we come in, we go out, but we never even look at the door on the other side of the landing. I recognize that that seems incredible, but that's the way it is.[22]

Another documentary adopted what had by then become a typical approach to the forgotten. *Les oubliés de la canicule*, directed by Sophie Lepault and Ibar Aibar in 2004, told the story of the forgotten by highlighting the life trajectories of a handful of the group. The film engaged in a deep humanization of the victims it profiled, and condemned both state and society for "abandoning" them. The film begins with the sun blazing on Paris's rooftops; a voice-over states flatly, in staccato phrases: "Summer 2003, you remember, the slaughter. In France, they died of heat. Fifteen thousand in a month. Families were on vacation, and late to claim their dead." The forgotten were "sixty-six lives, abandoned." They were "deaths that disturb us," and the funeral service at Thiais was an "expiatory ceremony" for the high state officials who were "absent during the crisis, but present on that day."[23]

Lepault and Aibar's profiles included five of the forgotten who figured in *Le Parisien*'s list. They included André Balateau, a construction worker who had abandoned his family in the 1960s, seeking work in Paris. He suffered a knee injury that rendered him unemployable, and lived on the street in the thirteenth arrondissement for years. Once he drew on state retirement funds, he rented a tiny room directly under the roof of a building on the Avenue d'Italie. According to his next-door neighbor and friend, during the heat wave, despite his disability, "No one came. No one." They included Pedro Santamaria, an exile from Franco's Spain who lived under the roof of a service building behind a luxury apartment building in the sixth, and Marie Antoinette de la Rochelle, a retired typist who lived in a tiny apartment in the seventeenth. And they included Micheline Bancalin, who lived under the care of the Petits Frères des Pauvres, a Catholic charity organization, and Daniel Serus, a militant unionist who had kept to himself in his retirement. Of these, Daniel had the only (relatively) happy ending. When his former coworkers and other union activists saw his name on the list, they pooled resources and claimed his body from the state in order to give him a decent burial. But his friends described their fallen comrade's death as "monstrous." It was "an alarm signal about the society we live in now. We are becoming dehumanized, completely dehumanized."[24]

These sentiments of collective guilt reached beyond journalism. The artist Constance Fulda, for example, was moved by the circumstances of the forgotten and spent months on a project honoring their predicament. She told me about how she had returned from her summer vacation and found a fax of *Le Parisien*'s list from a friend. She then conceived of a series of canvases that would pay respect to the group. As she described the project, she "had the idea to do 66 small works." Each is a square painting, about twelve by twelve inches. Most include a stylized representation of a gravestone that displays the name of one of the forgotten, with dates of birth and death, when available. The canvases are multimedia pieces, as much sculptures as paintings. The gravestones are cast in plaster, with the names and dates etched into them with a stylus and finished in gold leaf. The backgrounds are dappled colors: reds, greens, blues, and browns speckled with gold leaf and blacks. Each includes a slip of paper glued to the canvas, with the words "Homage to these 66 dead from the heat wave, parked in refrigerated trucks around Paris before ending up in complete abandonment" (figs. 8 and 9).

Fulda argued that the act of making each canvas brought each a bit of respect. Her hope was that each of the canvases would "find a welcoming home where a gaze would bring them a connection."[25] The first exhibition sought to emphasize the immensity of the heat wave and its capacity to expose France's social decompensation. Fulda mounted the images in November 2003 in the building that housed her studio, in the Rue de Vaugirard. The small canvases appeared in a long, dark, high-ceilinged hallway on cracked walls with flaking paint. The skylights above cast a diffused light on the images. Her intention was that the immensity of the walls would contrast with the small size of the paintings to emphasize the solitude of the victims. Fulda provided each visitor with a flashlight to view the images, composed of multiple colors with gold leaf, so that the light would glance off of the pigments' scintillations. Sales of the images benefited the charity group Morts de la Rue;

FIGURE 8. Constance Fulda, *Untitled*, 2003. Permission courtesy of Constance Fulda.

now the remaining canvases are in the group's possession.[26] Fulda also built a Japanese screen for the exhibit with the title *Brûlants oublis* (The Broiled Forgotten). One side of the screen shows a painting of a forest fire; the other is inscribed with names and dates of those who appeared on *Le Parisien*'s list, accompanied by a poem by André Duprat:

> 66 men and women
> 66 dead from the heat wave
> whose last voyage
> was in a refrigerated truck
> an identifying wandering
> in search of recognition
> France was hot[27]

These projects pointed to what their creators saw as a general public complicity in the deaths of the forgotten. The deaths resulted from the failures

FIGURE 9. Constance Fulda, *Untitled*, 2003. Permission courtesy of Constance Fulda.

of the state and of the community, of family and of neighborhood. As one author opined in *La Croix*, "How can we organize these existing solidarities, steady them when they waver? And not only during a heat wave or a cold snap? The answer belongs to the collectivity. . . . How can we reopen these doors that sometimes close? The answer is within us."[28] These works elicit sympathy for the fallen victims of the heat wave. They bring humanity to the fallen, and indicate that something as simple as knocking on a door, making a phone call, or acting in kindness constitutes the difference between abandonment and inclusion.

Yet other efforts to individuate the forgotten operate in a different rhetorical register, and indicate a more complex relationship between the forgotten and their memory. The centrist news magazine *Marianne*, for example, profiled three of the forgotten in an issue that ostensibly took state and society to task for their roles in the disaster. "After the summer's slaughter, shocking lies," the cover shouted at readers. Several articles took aim at mainstream representations of the disaster. One chastised the media and politicians for partisan portrayals of the disaster. Neither the thirty-five-hour week nor Mattei's polo shirt had anything to do with the heat wave's breathtaking mortality, it argued. Instead, the state's mismanagement of the hospital system and of elder care deserved the lion's share of the blame, as did a national obsession with youth and a culture of antiaging bias. But the centerpiece of the issue was its profiles of Marie France, Mihajlo Molerovic, and Philippe Heurteaux, three of the forgotten victims buried at Thiais. Each of these cases challenged conventional wisdom about the forgotten, and exposed a rhetoric that shifted responsibility for the disaster at least partially away from the French public and toward the victims themselves by highlighting the moral, personal, and emotional difficulties that contributed to their social marginalization.

Philippe Heurteaux was perhaps the quirkiest of the group. He had lived in an apartment that he had rented in the fifteenth arrondissement since 1954, and was a regular at the café on the corner. He was extremely thin and always dirty, "with the look of a hobo," according to the café's owner. He would show up every day to eat lunch and would buy drinks and food for others who would listen to his stories. The restaurant's regulars called him "Einstein," because, as the owner said, "he knew everything about everything and as soon as a conversation started, he would get involved in it. And he had to have the last word." According to local legend, Philippe's father was a chemist who had patented a substance that a drug company later developed. The patent yielded a small trust that was enough to support Philippe, who lived alone in the apartment with his cats. But Philippe's eccentricities extended to hoarding. He collected anything he found in the street, including furniture, broken lamps, and newspapers. His health degraded over time: he had developed Parkinson's disease, and his epilepsy worsened. On 14 August, the "millionaire hobo," as *Marianne* called him, died of heat stroke.[29]

When I visited Philippe's home, the custodian in the building next door confirmed this story of a paranoid, filthy, and possibly psychotic—if outgoing and generally friendly—figure. The story and the interview portrayed Philippe as a bizarre individual, one who had enmeshed himself in a social network around the café and the building but who had no real social bonds. While Philippe was charming in his own distinct way, Mihajlo Molerovic was a much more difficult subject. He lived not far from Philippe in the fifteenth in a public housing project in the Rue de Cronstadt that dated to the early twentieth century. Neighbors recalled the smell that emanated from the Serbian's apartment for four days after his death as they waited for the funeral services to pick up the body. They also recalled the constant shouting from the apartment before Molerovic's death. For ten years, he had shouted at, beaten, and insulted the woman who lived with him, an undocumented Serbian domestic servant. Molerovic was chronically ill, alcoholic, and disabled—rumors circulated around his building about what had caused his disabilities—but still found the strength to abuse Ilinka, who shared his studio out of necessity and out of fear of deportation. According to the article, the French state had provided for his lodging and sustenance, and he responded with contempt.[30]

But it was Marie France, the figure with whom this chapter began, who had the most shocking story. Heurteaux was by all accounts mentally ill; Molerovic was a violent drinker and a shut-in: neither their deaths nor their isolation was surprising. The aloof Marie France, with her close attention to her appearance and her fierce independence, was another story. Her formality and regal bearing suggested a privileged background. She had even at one point claimed a noble heritage, according to a neighbor. But she also had a difficult, even imperious personality, addressing her neighbors rudely at times. On 11 August, the building's caretaker went up to Marie's apartment to check on her early in the evening. She knocked on the door and heard a weak response. When she entered the "always impeccably clean studio," she found it in a shambles, and Marie seated naked on the toilet, "her cheek pressed against the wall, her eyes half-closed. She breathed weakly. A suffocating heat reigned." When the caretaker announced she was calling the hospital, Marie said harshly, "No, get out of here." The next day, the caretaker returned in the afternoon and found Marie splayed on the floor, dead.[31]

After her death, a different story about Marie France's life surfaced. According to city officials, much of her life had been a lie. She had never been married, and had never had a brother who fought in the Great War. As her upstairs neighbor told me, she had been a foundling, born in the same isolation in which she died. Her birth during the opening salvos of the Great War had prompted the orphanage to give her her patriotic name. Her refusal to let neighbors apply for social assistance on her behalf was a strategy to avoid their discovery of her humble origins. Neighbors also began to talk about her hypocritical moralizing: she said that it was "unclean" to sleep with a man, yet neighbors claimed to hear her entertaining young men whose favors she purchased.

The portraits in *Marianne* are complex, but they aim at a representation of the forgotten that is entirely different from the hopelessly lonesome and abandoned figures of Fulda's works. The individuals portrayed in *Marianne* are less the victims of uncaring family and society, and more the agents of their own marginalization. Their insanity, their violence, their dishonesty, are the roots of their social isolation. A range of other pieces in the press moved in the same direction. There was Jeanne S., a retired *gardienne or building caretaker,* who had voluntarily cut off all ties to her family, and who refused to say hello to anyone in the building or on the street. Neighbors knew better than to ask how she was doing—"She made it clear that it was none of our business," one said. There was Juan, a Spaniard who had fled Franco's regime and who delivered fruits and vegetables at Les Halles in its heyday. But after he had an affair, his wife had run back to Spain with his children, and cancer had left him a shell of his former self. He lived alone in a stifling, roach-infested apartment under the roof of a building in the first arrondissement.[32] And there was Michèle from La Réunion, only fifty years old, who lived in a single-room occupancy hotel in the eighteenth. Michèle had cut off all ties with her family and refused to seek any medical care, leaving the hotel only when she had to do so.[33]

These stories and others portrayed the forgotten as having sealed their own fates. They were, according to these inquiries, marginal characters who found it difficult to integrate into French society, but largely through circumstances of their own making. Even when stories portrayed some of these figures in a relatively sympathetic light, they described them as characters who colored the urban landscape without truly populating it. Conceived in this way, the forgotten did not threaten the social fabric in any real way, but instead they existed outside of it. They had withdrawn from society as a consequence of their actions—whether voluntarily or as a function of their erratic behavior, their madness, their addictions. The rhetorical power of such portrayals is to redistribute culpability and to direct blame toward the victims themselves. A story about an old woman who has aged in place while an uncaring society ignores her increasing isolation—a result of the deaths and distancing of those who inhabit her social circles—engenders deep sympathy and reflection on the collapse of solidarity. A story about a violent drinker who rejects all contact produces a different response. We imagine the latter as a sad case, but his isolation is both understandable and perhaps inevitable.

My research into the lives of the forgotten echoed the sentiments expressed in these investigations. My conversations with building custodians, neighbors, and shopkeepers often configured their lives in similar ways. As the stories of the victims' lives began with their deaths, the narratives that those who knew them shared with me pointed—even drove—toward an ending of indignity. Narratives in the press and in the memories of those who surrounded the forgotten suggest that this strategy is a mechanism through which many have come to cope with the tragedy of the forgotten and the discomfort their history produces. This reaction signals the important ways that the phenomenon of the forgotten sheds light on the imagination and memory of the disaster, as well as the limits of citizenship in contemporary France.

Aggregate and Anecdotal Life

In July, when I was about to leave, I offered to buy her some water and something to eat, and she only thought of her money. She had the means to feed herself. It was unfortunate like that. . . . She had the means to live and feed herself, but she couldn't not think of her money. . . . I said that I would pay for a few things for her and she could pay me. It wasn't serious, it was 25 or 26 euros. A little steak, some chicken, some yogurt, and some water, two liters. I brought all this up to her. . . . And she said that she only had a 50-euro note and that she couldn't pay me. That was Friday, I left on Sunday, and when I came back, she was dead. And she had the money. I'm telling you this because there are few things more important than living.

INFORMANT ON SIMONE, d. 13 August, eleventh arrondissement

One of the central problems of this book is identical to that of the media treatments above. When a group assumes its importance primarily in the aggregate, what problems does the examination of that group's constituents pose? For the news media that investigated the forgotten in brief articles, the phenomenon of the forgotten constituted either an indictment of French egoism or an explanation of individual cases of social isolation. This book makes a different argument. It contends that while the forgotten represent an important collective phenomenon, their individual stories are equally significant. It is through the disaggregation of their narratives of isolation that we come to understand their places in local social worlds. The group is too small and too selective to offer extensive interpretive value for understanding global mortality during the disaster. Yet their stories can provide critical insight into a number of problems that characterized the heat wave. They can tell us about the nature of living in poverty and isolation in the contemporary city, about aging and loneliness, and about the memory and social imagination of the disaster. While many might dismiss the individual histories of the forgotten as anecdotal, I argue that their very status as anecdotal constitutes their significance. These stories put a human face on the catastrophe, and signal the limitations of official accounts of the disaster.

The number of the forgotten itself is open to question. *Le Parisien* listed sixty-six names on 2 September. After a few families and friends claimed bodies, the cemetery buried fifty-seven the next day.[34] Later in September, a genealogical firm took on the cases of eighty-six of those buried in Thiais for investigation into any family connections they might find. And a walk through the cemetery reveals well over a hundred bodies sealed into individual tombs of divisions 57 and 58 with death dates in the month of August 2003. The discrepancies are somewhat easy to explain. By the end of August, city officials had already given up on finding relatives for some of those who had died early in the month, so they did not include them in *Le Parisien*'s list. Some of those buried in the tombs were not heat wave victims, but those who died from other causes. The genealogical search included the fifty-seven buried on 3 September and a number of others whose filiation remained unclear, even though the cemetery had buried them earlier in August.

This project counts ninety-five of them as relevant. In order to include victims in the list, I limited them to those who lived in one of Paris's twenty arrondissements. Several residents of nearby Paris suburbs lie in the plots because they died in Paris hospitals. But to include some residents of the *banlieue* and not others would raise too many complications by opening the inquiry to potentially quite different local ecologies. I also restricted this project to those who died between 1 and 20 August. These were the limitations in most of the official reports on the heat wave, and they make for the most useful comparisons between the group of the forgotten and the global population who succumbed to the heat. Several of the subjects I include here have official death dates later than 20 August. But reporting officials can only assign a death date based on when they find the body in question. In many cases, a death notice indicates that the reporter found the body days or even weeks after death. I have included cases where the death notice unambiguously points to a death between 1 and 20 August while still assigning an official date after 20 August.

Of the ninety-five victims, I visited the addresses of ninety-three.[35] Two were homeless victims who died in Paris hospitals with no indication of where they spent their final days. Well over half of them (fifty-four) were men. This fact places the group at odds with global mortality during the heat wave, in which women were significantly overrepresented. Another significant departure from wider mortality patterns is the average age of the affected group. Women among the forgotten conformed to the general profile of risk. The youngest was 42, the oldest 102, and the average age among them was 77. As with the general population, about 80 percent of them were over 75. But men among the forgotten differed dramatically from the norm. The youngest was 41, the oldest 88, and the average age was 66. Roughly 80 percent of the men were under 75—nearly the opposite of the general population. And also in contrast with the circumstances of the wider population, where a third of victims died at home, 80 percent of the forgotten died where they lived.

Certain basic social conditions predisposed the forgotten to isolation. Twelve of them were married—although not to other forgotten victims. Some of these were estranged from their spouses. But a clear majority had either never married (thirty-nine) or were divorced or widowed (twenty-nine).[36] Many of them had roots in Paris. More than a third had been born there, and nine of them died in the same arrondissement in which they had been born. Another third were born in provincial France, and eleven came from former colonies or the French Overseas Departments and Territories (DOM-TOM).[37] A further eleven came to France from foreign countries. Two had completely unknown origins. But the geographic origins of the forgotten as a group suggest that immigration was not a major factor determining their isolation. A plurality of them had lived in Paris for much if not all of their lives, and their isolation resulted more from social factors than from their physical displacement.

As I argue in chapter 5, their departure from broader mortality patterns tells an important set of stories. The overrepresentation of men among the forgotten suggests—as Eric Klinenberg argues for Chicago—that men are perhaps somewhat likelier to live in social isolation than women.[38] And when combined with the narratives of their lives and deaths, the overrepresentation of the relatively young among the victims indicates a range of social factors—including addiction and disability—that pushed them to the margins and into a social space of vulnerability. Their stories force us to rethink what has become an automatic association between age and vulnerability to extreme heat. Yet it would be inappropriate to invest too much in these figures. The group's composition is highly biased from the outset. The predicament of their burials in a state of relegation necessarily means that as a group the forgotten are likelier to have lived in isolation than those whom family members claimed and buried or cremated with private funds. They can therefore perhaps represent marginalization in a highly specific urban environment, but cannot stand in for a general population of victims.

The process of assembling the social histories of the forgotten involved listening to the stories that those who surrounded them told. But by definition these stories were fragmentary. As isolated subjects, the forgotten had few if any contacts who could provide more than a few bare details about their lives. They had lived their lives largely in the shadows of contemporary Paris. They inhabited the cityscape as loners whose survival on the margins ensured their status as isolates. As homeless, they dodged the violence and disease that thinned the population of their homologues; as the elderly, they had aged into loneliness as their friends and loved ones died around them. As those whom the city forgot in an ecological disaster and in their deaths, they were also those whom the city forgot in the course of their lives. So gathering their stories was a difficult process. In many cases—when a new *gardienne* had taken over the building since the heat wave, when no neighbors were available or willing to talk—I could do little more than verify that the subject had lived and died in the building I investigated. But in a surprising number of cases, I found informants who willingly shared their perspectives on the victims and their lives. In most cases they shared their insight as anecdotes about the fallen and the forgotten—anecdotes that both echo the basic assumptions about the heat wave that prevailed in media and political narratives of the disaster, and reinforce the marginality and indignity that characterized the forgotten victims' lives and deaths.

The anecdote has a long and important history as both a literary form and empirical evidence for ethnographers, historians, and literary critics. The anthropologist Clifford Geertz made extensive use of anecdotes as tiny encapsulations of culture. Sometimes, as in his essay on "thick description," the anecdote serves as a fragment apparently chosen at random from his field notes that has illustrative power. In other cases, as in his famous essay on the Balinese cockfight, an anecdote—in this case, the author and his wife narrowly escaping arrest as they flee a cockfight invaded by police—provides the entire frame for the essay.[39] The alleged excerpt from his notes that provides an entrée into "thick description" is for Geertz "a note in a bottle": it provides a glimpse into a past and foreign culture, which it is the ethnographer's job to situate. "The ethnographer," Geertz tells us, "'inscribes' social discourse; *he writes it down.* In so doing, he turns it from a passing event, which exists only in its own moment of occurrence, into an account, which exists in its inscriptions and can be reconsulted." But he (or she) inscribes through interpretation, not merely as a transcriber of the voices of the past.[40] Even when an anecdote describes something real, it is still a fabrication, a story that is arranged toward a particular end.[41]

For the literary critics Catherine Gallagher, Stephen Greenblatt, and Joel Fineman, the anecdote operates as a particular form because of its relationship to truth. It is essential to the scholar's credibility that the brief narrative reflect a lived experience: it cannot be pure fiction. For Fineman, the anecdote is a literary form that "exceeds its literary status" by virtue of its "referential access to the real."[42] Unlike an excerpt from a novel, it contains, Greenblatt argues, "a touch of the real" that is the source of its apparently "raw" rhetorical power.[43] Greenblatt and Gallagher build on this idea in their insistence that the anecdote has an almost unique capacity not merely to illustrate historical narratives, but also to rupture them by producing what they call "counterhistory" through its status as an interruption in a larger narrative.[44] Gallagher and Greenblatt refer here to Geertz, but also to such canonical figures in the production of counterhistories as E. P. Thompson, Raymond Williams, and Michel Foucault, all of whom use the anecdote both as evidence to advance their arguments but also in order to challenge dominant historical narratives.

Gallagher and Greenblatt make specific reference to an obscure essay by Foucault, "The Lives of Infamous Men," which appeared in 1977.[45] It is the introduction to a collection of texts that Foucault gathered from *lettres de cachet*, police records, and royal inquests. These texts are short snippets of narratives of miserable lives, which Foucault calls "an anthology of existences."[46] They reflect the encounter of marginal figures and state power in the late seventeenth and early eighteenth centuries, power that Foucault already sees as diffuse and capillary in its deployment. These anecdotes emerge from local petitions to the sovereign imploring the intervention of his absolute authority into often petty matters. Foucault argues that in his early works he sought to explicate such anecdotes in order "to know why it suddenly became so important in a society like ours to 'stifle' (like one suffocates a cry, a fire, or an animal)" a marginal figure: "I sought the reason for which we wanted so zealously to prevent these poor souls from wandering down unknown paths."[47] But toward the end of his career, he sought to collect these stories in order to let them stand for themselves. It is his criteria for inclusion that are most relevant to my intentions here. He established for himself some "simple rules" for including these anecdotes:

— They must belong to characters who actually existed;
— Their existences must have been both obscure and unfortunate;
— They must have been recounted in a few pages, or even better, in a few sentences, as briefly as possible;
— These narratives must not simply be strange or pathetic anecdotes, but in one way or another (because they were complaints, denunciations, orders, or reports) were truly part of the minute history of these existences, of their misfortune, of their rage, or of their uncertain madness;
— And the shock of these words and these lives must bear for us still a certain combined effect of beauty and fright.[48]

My intentions are different from Foucault's or Geertz's, but they draw their inspiration from both of their styles. I recount spoken rather than written anecdotes in order to inscribe them not necessarily with meaning, but instead as a means of understanding the enormous complexity of the problem of the forgotten. And I am interested in documenting the encounter of marginal figures with a broader social discourse—of the excluded with the languages of inclusion that they violated through their isolated lives and deaths. The brief narratives that appear in this chapter encapsulate in memory the obscure and unfortunate lives of the forgotten. They are stories that bear within them the diverse and always uncertain factors that shaped the isolation of the forgotten, and that strike me, at least, if not with "beauty and fright," then at least with the poignancy and sorrow that characterize this form of anecdotal life.

> *He had a friend, a bourgeois lady, and one day she called me and said I don't want him to come see me any more because he is letting himself die. I called the paramedics several times on him, but they said that there's nothing they could do. I could bring him something to eat, but I couldn't do anything about his health. And in the course of time, even though I had contact with him once a day, I saw him more and more diminished. And toward the end . . . he wore his pants that he had shat all over, and I said, "That smells bad!" I said, "Monsieur C., open the window, open the window." And he said, "No, no." And he never opened it. Toward the end, in the last two or three months, he was a bit lost I think.*
>
> INFORMANT ON ROGER, d. 8 August, thirteenth arrondissement

The excerpts from my field notes that appear throughout this chapter are anything but random. I have chosen them because of their brevity, their richness, and their emotional content. They convey sadness, disgust, and anger, often in equal measure. But even though I have selected these examples as standing out for the powerful images they conjure, I also see them as typical of the sentiments I encountered in my fieldwork. Informants were quick to point out the ways in which the forgotten had pushed themselves to the margins. Sometimes they offered descriptive facts. Jean-Pierre, for example, was fifty-six when he collapsed in the street outside his apartment in Paris's tenth arrondissement on 11 August; he died at the Hôpital Lariboisière later that day. According to a neighbor who pointed out where he collapsed, Jean-Pierre was enormously overweight and had a significant drinking problem. His only contacts were those who happened to be at the same bar at the same time. His alcoholism had severed him from a broader social integration; dehydration, a result of drinking only beer during the heat wave, was the immediate cause of death. Jeanne, who was eighty-six when she died in her tiny *chambre de bonne* on the eighth floor of a building in the twelfth arrondissement, was unapologetically rude to everyone she encountered. Neighbors tried to say hello, according to the building's current *gardienne* and shopkeepers in the area, but she simply glared at them, rejecting any social advances. As the *gardienne* of her building until her retirement, she could have had many connections among her neighbors. But her disposition was such that no one was surprised that she had died alone.

But more often my informants moved beyond descriptions and employed anecdotes to convey a similar message. Brief stories about the lives of their neighbors—or even a series of brief stories—went beyond mere description with their rhetorical power. The stories these informants told brought to the fore all the ambiguities and complexities that surrounded the lives and deaths of the forgotten. They elicited the tragic dimension of each story—Simone, whose frugality led to her death; Marie, whose shame about her origins entangled her in a web of deceptions with her neighbors and prohibited her from getting the help she needed; Roger, who despite his failing health and failing bowels refused all offers of aid—but they also displaced the responsibility for death from those who surrounded the victims in the direction of the victims themselves. They portrayed the victims as difficult subjects whose alienation from the community was self-induced; they were at best figures who had tried the patience of those who had surrounded them for decades.

In these cases, stories, rather than descriptions, conveyed almost allegorically the ways in which their deaths were—at least for the storytellers—both symptomatic of and the consequence of their marginal status rather than a collapse of solidarity. Many of the anecdotes indicated exasperation with the subjects. Claude and Marceline, for example, embodied the frustration of French entrepreneurs with the welfare state and its beneficiaries:

> *[The landlord] had been dying to get rid of them, because they only paid 80, 70 euros a month, because they were under the law of 1948 [a French rent-control law]. They'd been here since just before the war. And they had been living here for fifty years so they had frozen rent. Monsieur X wanted to get rid of them, he had tried to expulse them several times. It never worked, because they were protected by the law.*

Combined with stories about Claude's paranoia and his anger, characterizations of his lifetime of exploiting the system at the landlord's expense indicated a trying personality who had perhaps exhausted sympathy among his neighbors long ago.

At other times, the anecdotes suggested that the victims themselves had given up on life. Roger, for example, was losing friendships because "he was letting himself die." Marie, on the other hand, indicated her impending death in other ways:

She had a plant on the landing, a beautiful hibiscus that was dying. She couldn't take care of it any more. So one day, when she went out shopping, I got my courage up and said, "Marie France, would you like me to take care of your plant? Repot it with some new soil?" She said to me, "Yes, yes, yes! I'd like that a lot, I'll pay you." I said that wasn't what I was worried about; I cared about the plant that was dying. And as she didn't have a child, didn't have any pets . . . it would be too bad if it died, she would have nothing on the landing to take care of, because that's the only reason she ever came outside, to go to the landing to take care of her plant. Psychologically, for her it was important. . . . So I took care of her plant. I got a new pot, I replanted it. And there was another little woman who lived on the seventh floor who came out of her apartment and said, "Oh, you're taking care of the plant, that's great, because I saw that it was dying" . . . and then she said, "Why hasn't Marie come home from her shopping?" And later that day we found out that she had fallen in the street and broken her hip.

The dying plant in the hallway was a symptom of Marie's gradual slide into death. And the broken hip sealed her fate. It destroyed what was left of her mobility, but also retrospectively signaled to her neighbors the extent to which her deceptions had isolated her from the community. As they sought to obtain domestic assistance for Marie through the social welfare system, Marie appeared receptive. But when she found that she would have to share the details of her private life with those who surrounded her, she suddenly rejected their offers, as they threatened to expose her darkest secrets:

There was an inquiry into any family when she died. This person had always told us that there was a brother who died in the war of 1914, who was never found, but she had told us lies the entire time. And we found out that it had all been lies once the inquiry into her estate was made. We found out that she had no father, like she had told us, no mother, like she had told us, no brother, no one. But when she was alive, we had no idea, we thought that was the truth. . . . She invented a life for herself that she never possessed.

These stories performed an important rhetorical function. They were expiatory narratives that configured the forgotten as agents of their marginalization. The storytellers signaled the ways in which they had reached out to the victims: "I was always nice to him"; "I got my courage up and said, 'Marie France, would you like me to take care of your plant?'"; "I brought him something to eat, but he never responded to the interphone"; "I said 'Monsieur C., open the window, open the window.' And he said, 'No, no'." And in every case, the effort met with rejection. Through their troubled personalities, their afflictions, their deceptions, the victims had isolated themselves from a community of urban sociability and into a community of vulnerability.

Abandonment in a State of Disaster

What do we mean by the concept of "abandonment" or "forgetting" in the context of the heat wave? What does the status of "unclaimed" signal in relation to the bodies that crushed public authorities during the heat wave? What rhetorical purposes did these terms serve? For the filmmaker Danièle Alet, who directed a documentary on the "forgotten" in 2004, the notion of "abandonment" was a dramatic oversimplification of the story of death in isolation during the heat wave. It was, she told me, an exculpatory narrative that served a critical purpose for the media and the state in the management of the disaster. She conceded that abandonment was one possible truth of the disaster that sometimes captured the reality of a given death. But the media version was only one of many possible truths of the disaster, a black-and-white story line that failed to represent the circumstances of most of those who died during the crisis. In the heat wave, "everything is grey," she said.

The terms "abandoned" and even "unclaimed" are laden; "forgotten" somewhat less so. The former terms conjure the notion of responsibility. If a body is abandoned, then there is an actor who committed the act of abandonment. Far beyond negligence, abandonment signals a deliberate and even cruel act. "Unclaimed" is more passive, but operates in a similar register. A description of the bodies in the warehouses at Rungis or in refrigerated trucks at Ivry as "unclaimed" (*non réclamés*) implies negligence: it indicates that there is someone, somewhere, who has failed to fulfill an obligation to the deceased and to the state, which is now forced to manage publicly a death that should be a private matter. Both abandonment and a failure to claim imply the existence of direct social or kinship ties between the victim and others who have actively rejected a mandate to care for the victim. "Forgetting" has a wider scope. In labeling the heat wave's victims the "forgotten," the press described a broad social phenomenon in which certain actors assume a degree of anonymity and invisibility in everyday life. They are those who surround us, who live in our buildings, whom we see regularly or merely know about, but about whom we only rarely think or care. While the term "abandoned" targets principally the families of the victims, then—the only ones who have a legal right or obligation to "claim" a body and to alleviate the state's burden of managing its disposal—the term "forgotten" serves as an indictment of an uncaring society at large.

Yet in many cases neither descriptor is adequate, as Marie France's death shows. Never married and with no children, there was no one to abandon her during the heat wave: there was no easily identifiable party responsible for her care or well-being. Nor was she "forgotten." While her neighbors were away on vacation, the building's *gardienne* told me that she had looked in on Marie several times during the heat wave.[49] Marie was remembered rather than forgotten, and according to the *gardienne*, had violently rebuffed offers of help.

Alet's film, like many other interventions into the problem of the forgotten, tells the story of a tragic collective by reference to a few individual stories. Like Lepault and Aibar, she details the life of André Balateau, who one day left for work and abandoned his family for decades. She discusses the case of Patricia, a clearly disturbed woman who had walked away from her family years before her death in a wretched residential hotel in Paris. She profiles August, an Austrian artist who had one day vanished from his children's lives, dying deeply in debt in a one-room apartment surrounded by thousands of his works. In each of these cases, "abandonment" is an ambiguous descriptor. It pertains to the families more accurately than to the victims. As André's children tell Alet, "*We* are the abandoned ones here."[50]

Alet's film corroborates a powerful ambivalence that I often found in my research. The press had brutally indicted the families of the abandoned victims, blaming them for an inhuman callousness that had constituted the ultimate foundation for the heat wave disaster. These families pointed out the difficulties that their loved ones presented. Where *La Croix* argued that some two-thirds of the forgotten had had families who should have managed or even prevented their deaths, these families—often living in other countries—indicated that it had been years or even decades since the victims had been in their lives. The victims, nearly all the neighbors and *gardiennes* whom I interviewed claimed, had produced their own fates.

Yet Alet's film also brings with it a sensitivity that I found in few of my interviews. The family members she documents show a wrenching grief upon learning the details of their loved ones' lives. The stories of the forgotten destroy them. Their distance from these figures is completely understandable. Their mothers, their fathers, their siblings, had disappeared from their lives without a trace long ago. But they are wracked with guilt when they discover their loved ones' lives—and not merely their deaths—in abject isolation. They are horrified at what they find, largely because, despite the deeply troubled histories of their relationships, they still recognize the humanity of the departed and the tragedy of their life circumstances.

One of the more difficult problems of conducting my research into these social histories was its emotional content. As a historian I am more accustomed to the archive or the library and to the details of lives that passed decades ago into a written record. Most of the sources I have used in other projects detail snippets of evidence about a broad range of phenomena. They often contain emotional elements. And interviews I have conducted have often been passionate. But none have had the immediacy of those I conducted in the course of this project. I had many successful interviews and site visits while doing my fieldwork. But every success entailed the discovery of a tragic narrative that encompassed more than a horrific death. There are no happy endings in this volume, and as Marie France's and many other cases show, there were precious few happy beginnings either.

But what I found most intriguing in this process was the discovery of the way in which the stories my informants shared about the forgotten constitute memories of marginalization. Death in isolation does not normally prompt soul-searching. If a troubled and troubling figure with a complicated life history and no apparently immediate family dies alone in a given building, with no one to manage his or her affairs or to claim the body, we are unlikely to blame the community. The case is sad, perhaps even tragic, but such things happen. We cannot expect every neighbor, every *gardienne* to come to the rescue of everyone in isolation.

But the circumstances of the heat wave are different as a function of the scale of this problem, and they indicate the ways in which both the individual and aggregate histories of the forgotten provide such a critical window into the disaster. These stories indicate the inadequacy of simple narratives to encompass the scope of the disaster. Many of the cases I studied involved what appeared to neighbors as scandalous abandonment: situations in which family pretended not to know the victim in order to avoid funeral costs, and then rushed to seek their inheritances from the victim. I heard about victims whose families rushed to their apartments to take money from secret hiding places and then denied any relationship to the victim. And many cases involved admittedly difficult personalities: those whose addictions, instabilities, and petty cruelties would have made any reconciliation with family unlikely. But in most cases, something more ambiguous appears to have characterized the victim's life circumstances and relationships. In these cases, "abandoned" and even "forgotten" appear to be weak adjectives for describing such complicated lives and deaths.

What appeared instead in these cases is something that I am calling, for lack of a better expression, anecdotal life. The forgotten left few traces, and their lives assumed meaning only as a function of their deaths' collective nature and its symbolic value. Individually, their deaths would be sad stories. In the aggregate, they are a national tragedy, and one that in retrospect defines their biographies. It is now impossible to tell the story of Marie, of André, of Roger, or of Simone without beginning that story at its end. The indignity of their deaths has become the defining characteristic of their lives, the fact that frames their histories as a series of anecdotes leading inevitably toward that end.

In a 1986 article titled "L'illusion biographique," the sociologist Pierre Bourdieu argued that biography—or the story of a life—is an inherently fictional narrative process. As with Geertz's anecdotes, any life that is "organized like a history" must "sacrifice" its richness to a "rhetorical illusion," one that is directed from a specific beginning toward a specific end, both chronological and rhetorical. Biography ascribes meaning to life by linking it only to certain specific anchors in a social context. But, Bourdieu notes, making sense of a life without its insertion in a broader social geography is "as absurd as trying to make sense of a metro line without considering the structure of the network." Instead, life can only become sensible when understood as a series not of events, but of "placements and displacements in social space."[51]

If biography is itself a rhetorical construction, the anecdotal representation of life is among its more exaggerated forms. As with Foucault's "Lives of Infamous Men"—"recounted in a few pages, or even better, in a few sentences, as briefly as possible"—the stories my informants told about the lives of the "obscure and unfortunate" were "truly part of the minute history of these existences, of their misfortune." But there is also what Bourdieu calls a "logical order" to these short narratives, driving the life course in a specific direction that aims at an "artificial creation of meaning." As a technique, anecdotal life divorces the experiences of the forgotten from a wider social matrix, so that they become metro lines without connections, aiming only at the solitary terminus of life.

Anecdotal life indicates one way that a community might come to terms with a disaster by displacing culpability. These stories have a redemptive function: they cast the storytellers in a positive light by signaling their interventions in the lives of the forgotten and the obstacles they faced in the process. But they also operate at a deeper level. These narratives—and especially their insistence—constitute a fracture in the status of citizenship. Each of the figures I have discussed in this chapter was technically an enfranchised subject: they were vested political, civil, and social citizens. And yet the life of a figure such as Marie France highlights the limits of citizenship and its protections. Marie France's citizenship consisted entirely of her bureaucratic relationship to the state—indeed, that was her sole relationship at birth and at death. The state witnessed her birth and even named her; the state managed her death and even buried her. Yet ironically it was this relationship (and Marie's shame that surrounded it) that prevented her from claiming the assistance from the state that could well have preserved her life, a paradox that a series of anecdotes—a dying hibiscus, a broken hip, a "life she never possessed"—illuminates more fully than a mere biographical description. The most important point is perhaps that these anecdotes are closed narratives that preclude any alternative to a death in isolation. They therefore contain a final, marginalizing violence that suggests the limits of social citizenship in the contemporary welfare state. The anecdotes suggest problems that pertain to these individual lives—those of elderly or addicted or marginal isolates—but they draw their power from wider referents—those of old age and addiction and marginalization themselves. Where Alet's film emphasizes the humanity of both the troubled figures she profiles and the families who allegedly abandoned them, the constitution of victim biographies as anecdotal life that I have sketched here draws on such an imagination of marginalization to simplify the enormous complexities of the urban social landscape that the heat wave revealed by dehumanizing both the victims and any possible relationship with them.

3

Place Matters: Mortality, Space, and Urban Form

In August 2003, some of Paris's most glamorous facades concealed the city's most miserable deaths. Seventy-nine-year-old Pedro, for example, died on 12 August in his room in the Avenue de l'Observatoire, in a building situated adjacent to the Jardin du Luxembourg. Two days later, eighty-eight-year-old Louise died in a pristine nineteenth-century building in the Rue Ambroise Thomas. In the Avenue de Friedland, within sight of the Arc de Triomphe and not far from the Galeries Lafayette department store, authorities found the body of the eighty-year-old Paulette on 20 August. Similar deaths took place in far more marginal sites. Patricia and Michelle, forty-three and fifty, respectively, died within days of each other in dismal residential hotels in the eighteenth. And no one knows when Claude, 77, and Marcelline, 102, died in their coal-heated, rent-controlled apartment in the twelfth. It was only when neighbors complained to police about the smell that pervaded the building that authorities noted the deaths on 17 August, indicating that the bodies had been there, in bed together, for about a week.

Most of these victims lived nowhere near one another. Yet they died within days of one another, and were buried within feet of one another. This chapter draws on the social histories of these victims to address the links between place and vulnerability in the urban environment, as well as the historical particularities of those connections. The heat wave drew attention to a number of critical ruptures in the French state, including the failures of emergency response, a collapse in social solidarity, and the increasing isolation of the elderly, who constituted the majority of the disaster's victims. But the disaster elicited far less debate about housing and the geography of vulnerability. This chapter focuses on this problem with particular attention to a population that, while technically "housed," lives on the precipice of homelessness. Investigating the everyday lives and deaths of these victims reveals the critical importance of the geographic dimensions of risk in the urban landscape, as well as the historical development of the place-based vulnerabilities that the disaster cast in high relief.

Place matters to this chapter in at least two ways. First, disaster is a decidedly local phenomenon. "Global" climate change, for example, has local effects: disappearing islands, the Arctic Sea's encroachment on Inuit settlements, the incremental expansion of malarial mosquito ecologies, the desertification of the Sahel. Disaster on a smaller scale is also place based. A hurricane out to sea is a meteorological event. The same hurricane can make landfall with dramatically different outcomes, such as the near-complete destruction of the Lower 9th Ward in New Orleans and the relative sparing of the nearby affluent suburb of Metairie during Hurricane Katrina in 2005. In the case of the 2003 heat wave, place was a critical variable in determining risk. Paris experienced what epidemiologists call an excess mortality rate of 143 percent during the disaster, meaning that the city witnessed nearly two and a half times the normal August death rate during the disaster; despite higher average temperatures, Marseille, by contrast, experienced an excess mortality rate of only 25 percent, or about one and a quarter times the mean.[1] As this chapter argues, place-based vulnerability depended not only on microclimates, but also on factors linked to a broader human ecology, such as the intersection of architectural forms and patterns of urban sociability.

Following on this notion, I argue that place matters critically to the writing of the history of disaster as both setting and source. To paraphrase Pierre Bourdieu in his 1993 essay "Effets de lieu," history not only happens, but also "takes place": events and experiences materialize in a particular geographic context.[2] In Bourdieu's geography, place is both "physical" and "social": a site we can fix on a map but also one that denotes a precise social position. The overlapping of physical and social space offers a critical perspective on both the inegalitarian dimensions of disaster and its historical contingency. Although disasters such as heat waves, droughts, floods, famines, and tsunamis find their origins in global phenomena beyond individual human control, their effects depend almost entirely on the local human environments with which they collide.[3] Katrina did not strike poor neighborhoods and miss rich ones: instead, historically particular social, political, and economic factors made certain sites vulnerable and others resilient.

A similar dynamic operated during the heat wave in Paris. When researching the history of the 2003 heat wave, my visits to the sites where individual victims died—victims such as Pedro, Paulette, Patricia, and ninety others—allowed me to pinpoint the overlapping of physical and social space, the precise intersections of an urban geography of risk and the socioeconomic dimensions of extreme marginality. What emerged most saliently was that although the forgotten lived and died in virtually every neighborhood in the city, they shared a social geography of poverty and unsatisfactory housing. In particular, many of these forgotten victims lived in *chambres de bonne*, the former domestic servants' quarters under the roofs of many of Paris's buildings that are a legacy of Baron Georges Eugène Haussmann's reconstruction of the city in the nineteenth century. These sites were not merely settings of or contexts for disaster. They were also agents of mortality, which constitute important historical sources in their own right. The forgotten victims of the heat wave left few traces; some of the most significant windows into their lives are the physical environments in which they lived and died. An environmental history of disaster—and particularly, an ethnographically informed history of a contemporary disaster—must account for the sites in which that disaster "took place," and engage with those sites as critical sources that allow us to map this intersection of social and physical spaces of vulnerability. The *chambres de bonne* and other forms of marginal housing are an artifact of the past that attests to a lingering inequality in the organization of contemporary urban space; during the heat wave, these sites became a critical hazard for many of the urban poor. A reading of these physical and social spaces helps to illuminate the microgeographies of vulnerability in Paris during the heat wave, indicating the ways in which risk and resilience are site dependent. The chapter's larger goal is to provoke a conversation about the ways in which a historical sensibility is essential for thinking about contemporary disasters and their landscapes of vulnerability at the local level. The cases outlined here provide a qualitative complement to standard geographic and epidemiological analyses of risk by providing insight into particular circumstances of vulnerability. They offer a glimpse into the everyday predicaments of violence and vulnerability that attend poverty and marginalization in postindustrial societies, and open a window into the effects of global phenomena on the ground—or, in this case, on the seventh floor.[4]

Mapping Vulnerability

In many ways, my efforts to explore the stories of the forgotten were geographic. I sought to plot a spatial distribution of their mortality and vulnerability. Death notices and phone directories enabled me to find the addresses of ninety-three of the ninety-five victims.[5] I spent months visiting their addresses by bicycle, by metro, and on foot and collected as many stories of these victims as possible through interviews with building concierges, neighbors, and local shopkeepers. While in some cases I was able to confirm only that the victim had indeed lived in the building and had died alone during the heat wave, in other cases I secured detailed descriptions of the victims and their social histories. Given the distribution of these deaths throughout Paris and the diverse circumstances that surrounded each of them, they provide a critical means of exploring the spatial dimensions of risk during the heat wave, and in turn, provide insight into the role of urbanization and urban form in structuring health risk in periods of both crisis and relative normalcy.

A map of the addresses of the heat wave's "forgotten" victims indicates that they died throughout the city, in neighborhoods rich and poor, north and south, east, west, and center (fig. 10). But there is one specific and critical set of overlapping spaces in which their deaths unfolded: a geography, both social and physical, of poverty and precarious life.[6] This was a space of economic marginalization operative throughout the city, as pronounced in lofty bourgeois enclaves as in poorer neighborhoods on the city's periphery. The physical sites of mortality are critical to the disaster's story: they serve as important sources for investigating both the experiences of these victims and the historical factors that structured their vulnerability. Written sources about the disaster are easy to find: medical and media reports are legion; transcripts of political hearings are a click away. But the victims remain mute, having left few sources behind. The places in which they lived and died are one of the few traces of their existence. Yet they are also sources that provide useful insight into the factors that exacerbated the likelihood of the victims' deaths in the disaster.

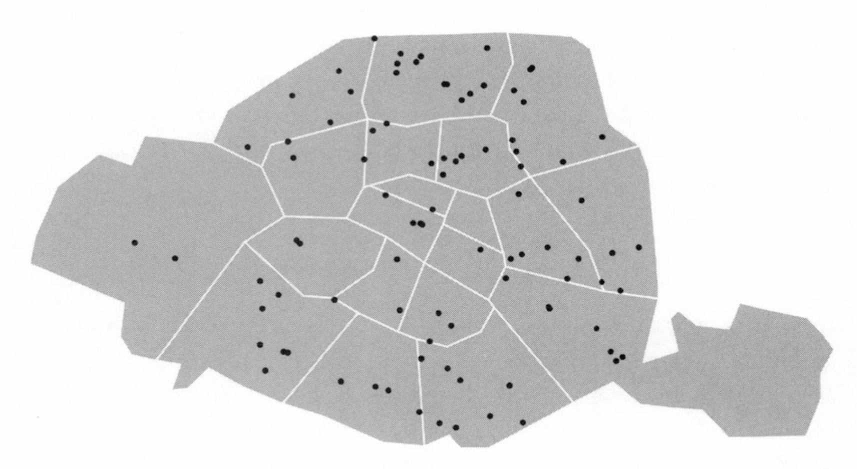

FIGURE 10. Addresses of the "forgotten" of the heat wave. Data collected by the author; map courtesy of the University of Wisconsin-Madison Cartographic Laboratory.

A handful of cases offer useful, although by no means unique, examples. At the southern end of the Jardin du Luxembourg, the Avenue de l'Observatoire begins with one of the more pleasant thoroughfares in Paris. Its median is a continuation of the city park, with statuary, children's playgrounds, benches, and tree-lined walkways set amid the green space stretching for two city blocks. On the west side of the avenue is the new site of the Paris campus of the École Normale d'Administration, the elite institution that has trained many of France's political leaders in the past century. On the east side are some of Paris's most luxurious apartment buildings. One of the addresses closest to the park stands out for the decorative tile work that lines the building's first-floor windows. From the exterior the building is by all appearances a well-appointed haven in one of the city center's most desirable districts.

When one enters the front door, however, it quickly becomes clear that it is actually two buildings in one. A marble-lined entryway gives access to the building concierge's apartment, the elevator, a wide staircase leading to the main building's apartments, and access to the courtyard. The courtyard provides access to the rear side of the building, which is accessible only by a serpentine and decrepit service staircase. This side of the building houses the *chambres de bonne*, the former quarters for the domestic staff who were once a common fixture in the grand apartments of such buildings. At present the rooms in these areas of many buildings—some owned by the main building's residents, some by independent owners and landlords—typically house students, poor immigrant families, and the elderly. On the top floor of this particular building, there are seven or eight of these apartments sharing a common toilet and sink at one end of the hall. When I visited the building on a cool early summer day in 2007, the temperature increased dramatically as I climbed the seven flights of stairs that led to the top floor (fig. 11). Although it was only 65°F or so outside, it was easily 85° or 90° under the roof. According to one resident I interviewed while visiting the property, students occupy most of the rooms. But for several decades leading up to the summer of 2003, four of the rooms were owned and occupied by Pedro. The child of refugees from the Spanish Civil War, Pedro was seventy-nine that summer. He had purchased the rooms and planned someday to join them into one larger apartment. He had few contacts: some family who had remained in Spain, for example, and a few acquaintances whom he knew from attending Mass at Saint Paul. But since his wife's death a decade earlier, the main building's residents who knew Pedro by sight reported that he had few visitors and only rarely left the apartment. On 12 August, when vacationers returned home to their apartments, they reported a horrendous smell. First responders found

FIGURE 11. Service staircase leading to Pedro's apartment, Avenue de l'Observatoire. Photo by the author.

Pedro's body at 10:00 p.m., and he became one of some 2,000 excess deaths recorded that day throughout France.[7]

Across the Seine in the eighth arrondissement, the Avenue de Friedland is one of the spokes of the "étoile," the star-shaped network of grand boulevards that radiate outward from the Arc de Triomphe. The Avenue de Friedland links the arch to the Boulevard Haussmann, the site of the Galeries Lafayette and Printemps department stores. The buildings here are classic examples of Haussmannian Paris: six- or seven-story buildings with the first floor dedicated to retail on a wide avenue originally established in the mid-nineteenth century. One is a solid dove-colored stone building with decorative iron railings in front of its apartments' tall French windows, like the thousands of others that give central Paris its homogeneous nineteenth-century character. Yet this building is also bifurcated. The elevator, which goes to the sixth floor, has a door in the rear as well as the front, and the upper stories have service apartments situated along a rear corridor. At the sixth floor sit these residents' mailboxes, separate from those of the main building's residents on the ground floor. These residents are fully segregated from those of the main building, aside from sharing an elevator. Their quarters share none of the luxuries enjoyed by the main building's residents. Conduit lines and loose wires run the length of the corridor. The individual rooms' doors are covered with chipped red paint, and are identified only by black-stenciled numbers on each door. There is a narrow, worn staircase that leads to these apartments and extends up a story beyond the elevator's reach into an attic space. This corridor, lined with mousetraps when I visited, is where Paulette died on 20 August (fig. 12). Eighty years old in 2003, she died in a tiny single room directly under the building's roof.

Not far away in the ninth arrondissement, Louise lived on the sixth floor of an impeccably maintained building in the Rue Ambroise Thomas (fig. 13). The building contains mostly apartments, but also housed a travel agent, a pharmacy, a suite of physicians' offices, and a dentist's office at the time of Louise's death in 2003. Louise had moved into the apartment with her husband over sixty years earlier, and had lived there alone since he had died decades earlier. They had never had children. According to the building's concierge, Louise lived her life in near-complete isolation: "She was the type

FIGURE 12. Hallway outside Paulette's apartment, Avenue de Friedland. Note the mousetraps on the shelf outside the door. Photo by the author.

FIGURE 13. Building in which Louise died, Rue Ambroise Thomas. Photo by the author.

who never wanted to bother anyone." Aside from a social worker, she had no contacts. When on 14 August the concierge knocked on the door, there was no answer, so she telephoned the *pompiers* (firefighters). They arrived and found Louise's body; based on its state of decomposition they estimated she had died at least several days earlier. Funeral workers were overwhelmed at this point in the crisis. Because of the backlog, Louise remained in her apartment for several days after the *pompiers* had found her, until the city finally transported her body to a refrigerated truck in Ivry.

Farther north, Patricia lived alone in a room on the top floor of a dingy hotel in the Rue Letort, located in a decaying section of the eighteenth arrondissement near the Porte de Clignancourt metro stop (fig. 14). The eighteenth is one of Paris's poorest arrondissements—according to 2008 census data only the nineteenth has a lower mean household income[8]—and the Rue Letort is in one of its most marginalized districts. Patricia had mysteriously left her family years before her death at age forty-four in 2003, having simply run away. By most accounts, she had been unstable mentally. She had no identifiable source of income, although she had consistently paid her

FIGURE 14. Residential hotel where Patricia died, Rue Letort. Photo by the author.

monthly rent for her top-floor room in the hotel for the five years she had lived there.

While they lived in different neighborhoods and represent wide-ranging social backgrounds, Pedro, Paulette, Louise, and Patricia shared important geographic and social circumstances that put them at inordinate risk. Not least of these was a vertical geography prescribed by poverty: each of these victims, along with many others, lived in rooms directly beneath the roofs of their buildings—many of which violated housing codes—where they baked to death during the unprecedented heat wave. Their deaths are inseparable from the poverty that pushed them into such precarious housing. Yet poverty alone is not enough to explain their deaths during the heat wave. For example, Paris's homeless population appeared to have experienced relatively few heat deaths in 2003.[9] Rather, the particular agency of heat waves preys on specific subsets of the poor, where poverty combines with place to exacerbate vulnerability to extreme heat.

A number of national and local agencies have explored the geography of the disaster's impact from multiple angles. The French meteorological service, for example, produced maps that showed concentrations of high temperatures throughout the nation, while epidemiological services such as the Institut de Veille Sanitaire (InVS) and Inserm generated maps that showed spatial distributions of the death toll.[10] Subsequent investigations aimed at more sophisticated analyses, dissecting global estimates of mortality and establishing profiles for those at greatest risk.[11] Data suggest that it was large urban agglomerations in central France that bore the lion's share of the mortality burden. These studies indicated that fully a third of excess mortality during the heat wave was concentrated in the Île-de-France (see fig. 15);[12] Paris itself, which represents about 3 percent of the French population, suffered over 7 percent of the mortality burden in the heat wave.[13]

There are strong environmental and social explanations for such differences in the mortality burden. Climatologists and epidemiologists have long attributed such differences to what they call the urban heat island effect, which links land-use patterns with heat concentrations. A preponderance of reflective surfaces, vehicle exhaust, ozone concentrations, tall buildings' capacity to stifle breezes, large concentrations of heat-producing sources, and other factors particular to urban environments create important temperature differences across landscapes, at times as much as twenty-two degrees Fahrenheit between downtown regions and surrounding rural areas.[14]

The heat island concept points to the coupling of human and natural systems as the space in which disaster unfolds. Human agency, in the form of the built environment, plays a role in shaping vulnerability to meteorological

TABLEAU III.2. : Répartition régionale des décès du 1er au 20 août

Régions	Nombre de décès observés (O)	Nombre de décès attendus (E)	Excès : O - E	Contribution à l'excès global	O / E
France métropolitaine	41621	26818,6	14802,4	100,0%	1,6
Alsace	1023	748,0	275,0	1,9%	1,4
Aquitaine	2191	1567,0	624,0	4,2%	1,4
Auvergne	1022	747,4	274,6	1,9%	1,4
Basse-Normandie	822	697,2	124,8	0,8%	1,2
Bourgogne	1477	885,1	591,9	4,0%	1,7
Bretagne	1855	1549,9	305,1	2,1%	1,2
Centre	2441	1203,4	1237,6	8,4%	2,0
Champagne-Ardenne	988	629,1	358,9	2,4%	1,6
Corse	191	143,4	47,6	0,3%	1,3
Franche-Comté	687	494,7	192,3	1,3%	1,4
Haute-Normandie	1066	764,3	301,7	2,0%	1,4
Ile-de-France	8506	3639,1	4866,9	32,9%	2,3
Languedoc-Roussillon	1536	1265,9	270,1	1,8%	1,2
Limousin	651	469,9	181,1	1,2%	1,4
Lorraine	1526	1066,4	459,6	3,1%	1,4
Midi-Pyrénées	1762	1324,5	437,5	3,0%	1,3
Nord-Pas-de-Calais	2175	1792,3	382,7	2,6%	1,2
Pays de Loire	2399	1430,9	968,1	6,5%	1,7
Picardie	1153	817,9	335,1	2,3%	1,4
Poitou Charente	1432	872,0	560,0	3,8%	1,6
PACA	3194	2375,1	818,9	5,5%	1,3
Rhône-Alpes	3524	2335,3	1188,7	8,0%	1,5

FIGURE 15. Regional distribution of deaths during the August 2003 heat wave. Source: Hémon and Jougla, *Surmortalité liée à la canicule d'août 2003—Rapport d'étape* (Paris: Inserm, 2003). Permission courtesy of Inserm CépiDc.

extremes beyond human control. Yet the heat island concept still privileges temperatures over social practices as the critical determinant of morbidity and mortality. An alternative explanation for high urban mortality during heat waves is more compelling for social scientists and humanists. This explanation is based on patterns of sociability in urban versus rural environments. The hypothesis is that social and community ties are stronger in villages than in cities. Yet this notion is based as much on a romanticization of village life as on any measurable reality. In a sense, both the urban heat island effect and the social explanation are new manifestations of the notion of the city as a pathological environment, an idea that itself dates at least to the late nineteenth century. Although early modern cities often struck visitors as noisome and disease ridden, expanding industrialization and its concomitant pollution, labor migration, and overcrowding engendered a new construction of the city as a locus of sickness in the popular European and American imagination, a space populated by teeming hordes and criminal behaviors, a dramatic contrast to a vision of nature as a salubrious site for restoration.[15]

As in the late nineteenth century, the contemporary city's microgeography of mortality has at least as much to do with local forms and disorders of urban sociability and habitation as it does with the city itself as a pathological mechanism. The key question is perhaps not *whether* cities show higher mortality rates than other environments, but instead *where* cities show concentrations of mortality. What particular environments and neighborhood structures coincide with higher risk of death during heat waves and other disasters? Such an analysis allows for a finer accounting of how to evaluate risk and vulnerability in urban communities by potentially exposing more specific factors that influence health outcomes. These factors might include building styles, neighborhood layouts, crime rates, concentrations of relative wealth and poverty, access to transportation, and proximity to hospitals and health care services among others.

Two key studies have produced models for this kind of endeavor. The first is Karen Smoyer's place-based analysis of heat waves in St. Louis, which signaled high risk for heat-related mortality in the poorest sections of the city. The critical factor exacerbating heat mortality was not the local heat load, but rather socioeconomic decline.[16] The second is Eric Klinenberg's study of the Chicago heat wave of 1995.[17] Klinenberg studied two adjacent neighborhoods that shared a constant heat load, but experienced dramatically different mortality levels. The study indicated that despite the roughly similar socioeconomic status of the two neighborhoods, the neighborhood with the higher mortality rate was marked by high rates of crime and social decline, while the other fostered a higher degree of sociability among its residents. It was the fear of leaving one's home that exacerbated risk in this case, Klinenberg argues: higher degrees of sociability encouraged at-risk populations in the "safer" neighborhood to leave their sweltering apartments and to seek cooler locations in order to beat the heat.

While a number of studies of the 2003 heat wave in Europe have paid close attention to the spatial dimensions of vulnerability, those generated by the Atelier Parisien de Santé Publique (APSP) stand out for their emphasis on mapping. The Paris mayor's office hired the APSP's staff to conduct a number of official studies of the distribution of mortality in the city. Several studies highlight the group's key findings.[18] Their maps plot excess mortality throughout the city, indicating important intersections of neighborhood and risk.[19] Of these maps, the most nuanced plots mortality by *quartier* (quarter) rather than by arrondissement (each arrondissement is divided into four smaller administrative units).[20] These *quartiers* more closely resemble neighborhoods than do arrondissements: they tend to be more economically, socially, and politically homogeneous than the larger units. Plotting mortality by *quartier* allows for a more sophisticated analysis of other data available for these units, such as socioeconomic status of local households and the median age of the local population. The study plotting mortality by *quartier* pointed to a specific nexus of mortality risk, in which increased age of the population and lower socioeconomic status correlated with high death rates during the crisis. The map also reinforces the importance of the heat island effect: some of the sites of the highest mortality are in the districts with high concentrations of *grands ensembles*, the massive concrete, steel, and glass high-rise structures that characterized major projects such as the Front de Seine in the fifteenth arrondissement and Les Olympiades in the thirteenth, both of which date to the 1970s.

These are sophisticated studies. They involve careful overlays of extremely precise quantitative data: plots of mortality by street address and economic indicators about per capita income by city block. They draw exhaustively on GIS data and involve careful digital modeling. Yet they are not field-based projects. They go no further than mapping a spatial correlation between death and economics. Such projects are essential to city planners and public health officials, as they reveal areas of risk at a glance. But they tell us little about the experiences of living in these vulnerable environments. They leave an important space for qualitative research that can begin to develop a richer understanding of the social and historical factors that condition vulnerability to disaster and the violence of everyday life on the margins.

Health, Hygienism, and the Making of the Modern City

Another map, this one dating from a century and a half before the heat wave, throws intriguing light on the heat wave's cartographic epidemiology (fig. 16). This map depicts mortality during the 1849 cholera epidemic. During the epidemic, some 20,000 Parisians died. This was despite extensive sanitation efforts implemented by public health officials in the aftermath of the 1832 epidemic, which claimed roughly the same number of victims. There are several interesting elements to the map. It shows a fascinating gradient of mortality. Deaths are thin on the ground in the northwest of the city, and are concentrated in the southeast of the city. What is perhaps most remarkable about the map is its stark parallel with a similar map of mortality during the heat wave (fig. 17). In contrast to the map of excess mortality by *quartier*, this map, produced by the same team, plots aggregate mortality during the heat wave, as does the 1849 map for the cholera epidemic. This is an important distinction. Maps showing districts with high excess death could indicate extremely high mortality during the crisis they depict, or they could indicate very low baseline mortality in the same district. By contrast, a map of aggregate mortality depicts strictly the number of deaths in a given district in a given period, with no control for average mortality levels.

The similarity is curious on several levels. First, it appears as if not much has changed in the intervening century and a half: death rates follow a smooth geographic gradient from few in the northwest to many in the southeast, following in both 1849 and 2003 a similar track from the high average household

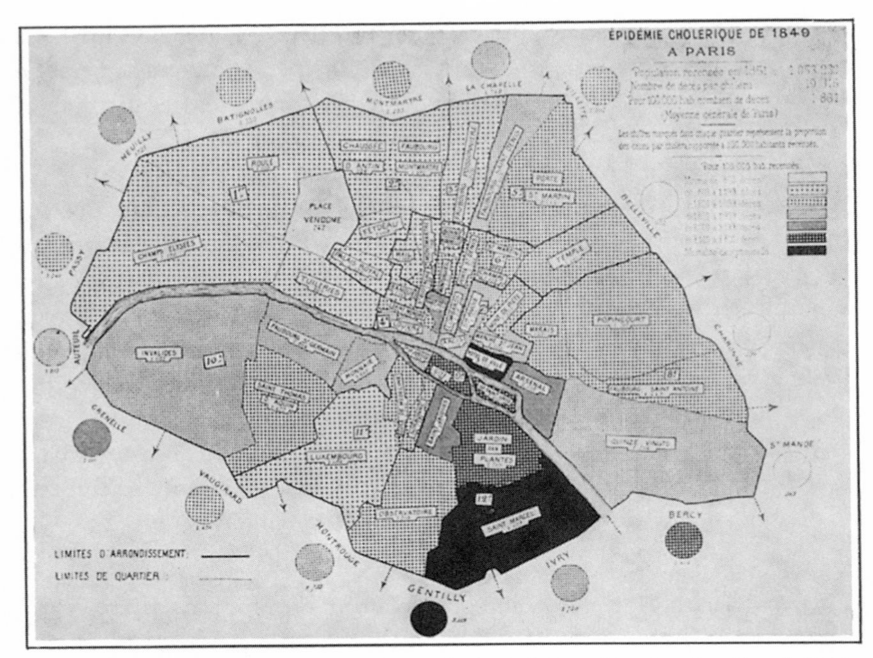

FIGURE 16. Mortality during the 1849 cholera epidemic, Paris. From Pierre Lavedan, *Nouvelle histoire de Paris: Histoire de l'urbanisme à Paris* (Paris: Hachette, 1975). Bibliothèque Nationale de France.

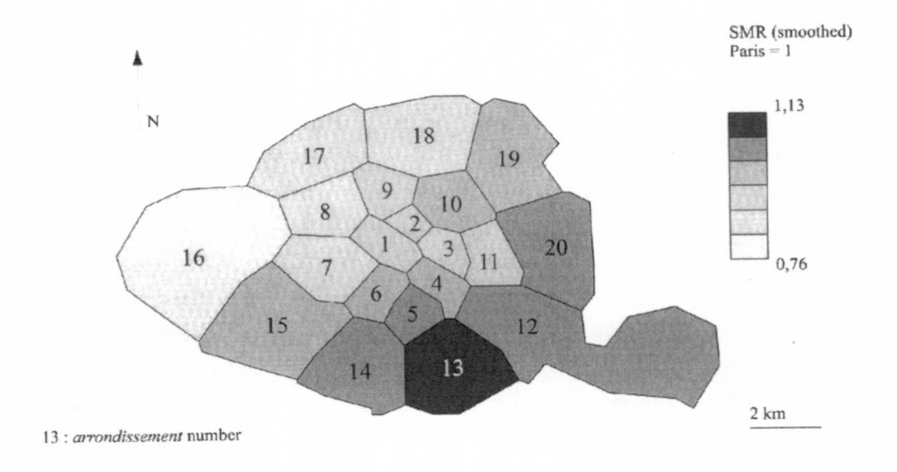

FIGURE 17. Aggregate mortality in Paris by arrondissement, 1–20 August 2003. Source: F. Canouï-Poitrine, E. Cadot, A. Spira, "Excess Deaths during the August 2003 Heat Wave in Paris, France," *Revue d'épidémiologie et de santé publique* 54 (2006): 127–35. © 2006 Elsevier Masson SAS. All rights reserved.

revenues to lower ones. The similarity also serves as a reminder that epidemiological projects like those of the APSP—mapping vulnerability; documenting intersections of land use and health concerns; exploring connections among diverse sources of data, such as vital statistics, census data, and economic indicators—have a deep history.

But perhaps the most interesting aspect of the 1849 map is its status as an artifact. It reflects a moment in which public health concerns originating in a moment of catastrophe proved to be enormously influential in shaping urban space. The Paris of today continues to show the influence of Baron Georges Eugène Haussmann, the prefect of the Seine charged by Napoleon III with rebuilding the city beginning in 1853. But that new shape of the city itself emerged from deep concern with differential mortality both in epidemic periods and under normal conditions. Among Haussmann's most lasting contributions was the reimagination of the city as a technological system that paralleled the body in its functions to maintain the health of both population and landscape; yet Haussmann's architectural legacy also contributed to a particular geography of risk during the heat wave. The concerns that prompted the drawing of the cholera map and the reconstruction of the city contributed to an effort to control sickness and health through urban planning. Yet Haussmann's remaking of the city—with an architecture that reinforced powerful economic inequalities in central Paris—contributed at least partially to exacerbating a vulnerability that the APSP's maps are unable to capture in aggregate form.

Intuition and anecdote had long fed assumptions about the relationships between health and wealth in France's capital (as well as in other European cities). By the 1820s a large body of newly available statistical data allowed policymakers to demonstrate these relationships conclusively. In the late 1820s, the physician and hygienist Louis-René Villermé pooled these sources of data in an investigation of mortality in Paris.[21] His central concern was the explanation of differential mortality rates in the capital: what factors explained the high death rates in some districts and low death rates in others? Villermé explored a range of factors that might have influenced mortality, according to contemporary medical knowledge. Proximity to the river, access to water, population density, architectural style, street layout, and wind patterns all came under consideration.[22] Yet none of these factors shed any consistent explanatory light on the project. The only consistent correlation Villermé found in his investigation was the overlap between low socioeconomic status and high mortality rates: poorer neighborhoods had higher death rates, and vice versa.

Accurate plotting of these demographic and economic data served useful purposes for local and national officials. In particular, such baseline information about life and death in the capital proved useful when cholera struck Paris in 1832. The epidemic spurred the establishment of a sanitarian commission that included Villermé.[23] The commission's studies of the epidemic (as well as studies of the 1849 epidemic) firmly placed public health concerns on the urban planning agenda, and such concerns were at the forefront of the new emperor's priorities when he charged Haussmann with rebuilding the capital.

Historians of Paris have cited a wide range of motives for Napoleon III's reconstruction of the city, including the emperor's ego and his concern with political security. Yet as David Barnes, Anthony Sutcliffe, and other historians have argued, public health concerns were at least as important: an effort to clean up the filth of the city and its concomitant diseases had characterized the project from the start.[24] Paris's population more than doubled in the mid-nineteenth century, growing from some 800,000 in the 1820s to more than two million in the 1870s. This congestion placed unsustainable pressure on the city's housing, water, and drainage infrastructure, and created a perfect ecology for diseases such as cholera, tuberculosis, typhus, and other killers.[25] Hygienist concerns and rhetoric were deeply entwined in the building project, which, as the geographer Matthew Gandy has noted, employed bodily metaphors to describe a functioning city as a "holistic" body.[26] The Bois de Vincennes and the Bois de Boulogne, at the city's eastern and western extremes, respectively, were the lungs of Paris, while its new fresh water systems and its main traffic arteries were its circulatory systems; the new sewers, among the most critical of these systems, were the city's bowels. Haussmann largely succeeded in his goal: cholera never returned to the city in a serious way (although TB remained a powerful scourge), and the British hygienist and social reformer Edwin Chadwick purportedly said to Napoleon III, "May it be said of you that you found Paris stinking and left it sweet."[27]

The remaking of Paris has generated an extensive historiography. Historians, art historians, urban historians and planners, architectural critics, and critics of modernity from the mid-nineteenth century onward have alternately praised Haussmann's vision and condemned his literal bulldozing of traditional forms. One of the most enduring traditions of this literature comes from Marxist critics who have taken aim at Haussmann's attacks on the poor. As David Jordan has noted, hygienic mapping projects pointed to zones of filth and decay in the center city that became targets for renewal: the bulldozing targeted precisely the slums whose staggering mortality levels Villermé had noted in the 1820s. Most of these neighborhoods were destroyed in their entirety. New construction, by contrast, involved for the most part the staid apartment buildings and retail architecture that remain central to the city's character. By the 1860s, painters such as Manet, Monet, and Degas captured in their images a new Paris of cafés, picture windows, and six- or seven-story buildings, which had begun to emerge from the rubble: a Paris of luxury and leisure was born out of the destruction of the city's poorest and most marginalized districts.[28] The displacement involved some 300,000 people, forced from miserable, squalid housing into homelessness.[29]

The mythology that emerges from this historiography of displacement is one of the literal marginalization of poverty in Paris: the establishment of a bourgeois city center ringed by an emerging proletarian "red belt" in the outer arrondissements and, eventually, the Paris *banlieue*.[30] This is true in an aggregate sense. Beginning in the mid-nineteenth century and extending to the present, central Paris has become a bourgeois entity. Yet while Haussmann's projects did displace poverty from central Paris, they failed to eradicate it. Indeed, one could argue that the destruction of the old city in the making of the new may well have increased economic inequality in the city center. In a famous illustration from the 1850s, Edmond Texier presented a cross-section of a "typical" pre-Haussmann Parisian building, which reveals a caricature of the city's vertical geographies of class (fig. 18). The ground floor is a working area, with an entryway, a kitchen, and a concierge's rooms. The first floor is an expansive bourgeois spread with elegant Restoration furniture, sumptuous draperies on the windows, and an elaborate chandelier hanging over a detailed marble fireplace. The second-floor apartment is a bit more modest: the ceiling is a bit lower, and the details of the apartment a bit less luxurious. On the third floor, the same space is divided in two, and the apartments are far more meager in their furnishings. On the fourth floor, vertical and horizontal space has shrunk dramatically. Now three apartments occupy the same square footage as one apartment on the first or second floor. All three apartments are directly under the building's roof, and those on the ends of the building have their space further compromised by the angles of the roofline.[31]

Several historians have drawn on these images to argue both for the inherent inequality and for the social solidarity of Parisian society in the pre-Haussmann era. Such buildings show a social stratification, in which poverty increases as one ascends the staircase. Yet such structures also indicate that rich and poor did live in the same buildings. Many romanticized this notion of solidarity, arguing that Haussmann's remaking of the city changed these configurations.[32] The destruction of slums created hundreds of thousands of homeless. But so did the destruction of many apartment buildings that conformed to Texier's caricature. Widening avenues and boulevards necessitated

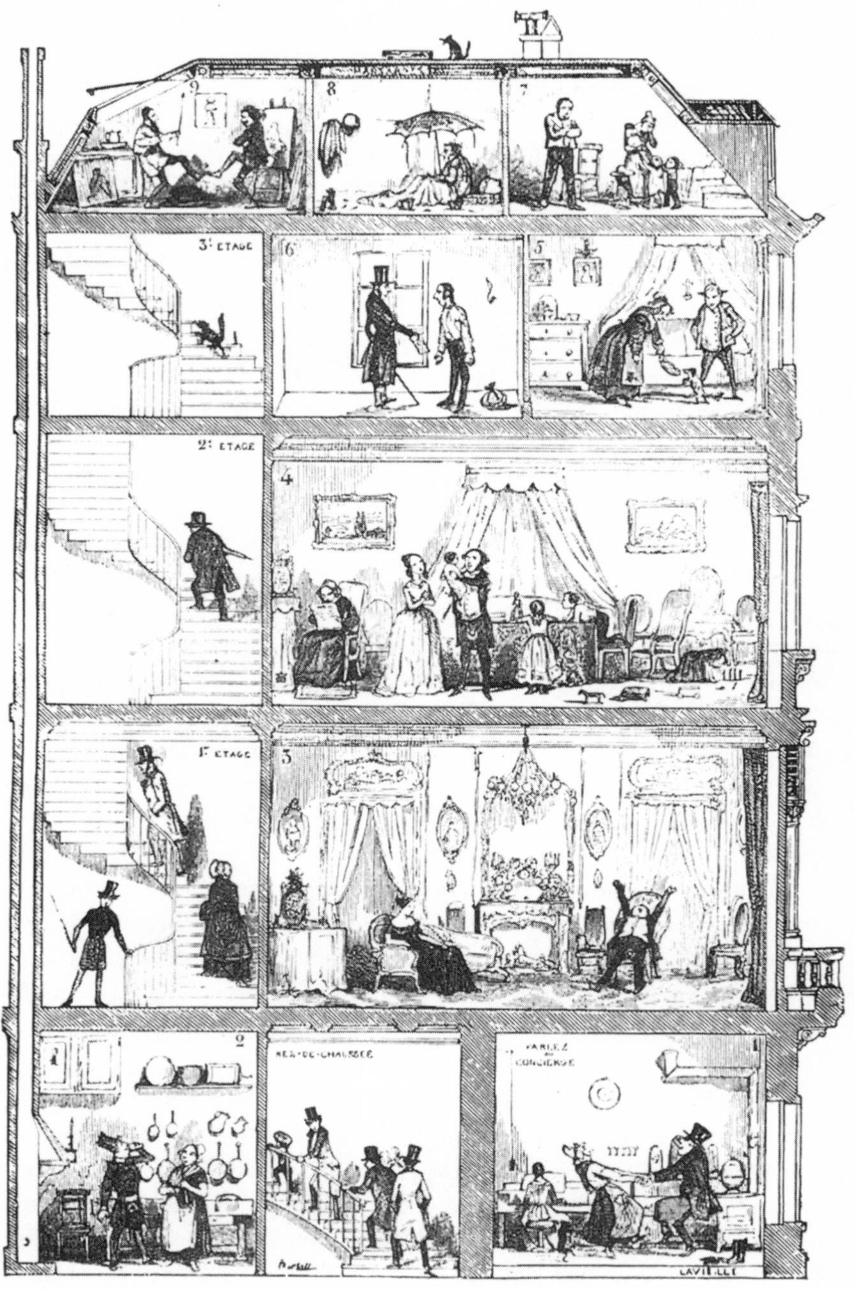

FIGURE 18. Edmond Texier, *Tableau de Paris.* From Pierre Lavedan, *Nouvelle histoire de Paris: Histoire de l'urbanisme à Paris* (Paris: Hachette, 1975). Bibliothèque Nationale de France.

demolition on an unprecedented scale. Developers recycled the rubble from these buildings and constructed entire new proletarian compounds on the eastern and southern outskirts of the city.[33] Those who could afford to do so moved to these new workers' cities; many others simply lived without shelter. But very few stayed in place. The rents in the new buildings that went up in place of workers' old residences were far too expensive for working-class wages to bear.[34]

This transformation created a new geography of class in the city that has remained essentially in place to the present: bourgeois enclaves in the center of the city and the western districts, proletarian strongholds in the east and the south. But this cartographic social division was far from uniform, in part owing to new building styles that prevailed in Haussmann's Paris. While the prefecture offered tax breaks and other incentives to encourage private building, it also imposed restrictions on builders' plans. Maximum building heights were fixed based on street width, leading to the six- or seven-story building that shapes the Paris skyline. Developers nearly always met this height limit as a means of maximizing rents. Other strategies for increasing revenue included giving the ground floor over to retail or other commercial outlets, which commanded a higher price per square foot, and using a mansard roof with dormers so as to include an extra story of rentable rooms directly under the roofline.[35] Building codes mandated only exterior building height, with no restrictions on the ceiling heights of individual floors or square footage of individual apartments. Therefore, while official statistics indicated that the number of individual dwellings expanded under Haussmann's plan, this expansion most often reflected the partitioning of individual apartments rather than a real accumulation of new properties.[36]

The result was the establishment of new quarters in residential buildings that dramatically increased population density on these "extra" upper stories, even as the number of total buildings within the city limits decreased. This architectural phenomenon points to the dual effect of Haussmann's projects in shaping a spatiality of class in Paris. Widespread demolition and the construction of inexpensive housing on the eastern borders of the city pushed Paris's artisanal workers to the margins, creating a social and political polarization of the city. But at the same time, new building styles also increased economic inequality in the city center. The poor who remained in central Paris and who occupied these new upper-story apartments were typically domestic servants who worked in the larger apartments below; housing in these conditions was often the bulk of their compensation. This phenomenon is the origin of naming such rooms *chambres de bonne*, or "maids' quarters." But as the historian Ann-Louise Shapiro has argued, this restructuring also meant the impossibility of restoring the "mixed-housing" model of the past, as workers were typically "too proud to inhabit servants' quarters."[37]

The *chambre de bonne* remains an important fixture of the Parisian landscape, yet it is virtually absent from the historiography of French urbanism, which has focused almost exclusively on postwar architectural developments such as the *grands ensembles* and habitations à loyer modéré (HLMs) of the *banlieue*. It is a critical site of romanticism: the starving artist's garret in a contemporary form. A number of glorifications of the 1968 revolutions begin with the author's reminiscence of living in a *chambre de bonne*, worn as a badge of radical credibility. Such rooms sheltered Holocaust survivors at great risk to their proprietors, and served as loci for both bourgeois sexual subjection of domestic servants and working-class women's resistance to that subjection.[38] More important, the *chambre de bonne* is a sign of the persistence of desperate poverty amid unimaginable wealth. These rooms are vestiges of a Dickensian social inequality from an era that preceded the development of the welfare state. Even in France, where the welfare state is a critical element of social and political citizenship, the *chambre de bonne* is a hidden sign of the state's limits of protection.

Mapping Vulnerability Redux: The Heat Wave's "Forgotten" Victims

As the heat wave of 2003 made all too clear, the *chambre de bonne* is also a clear sign of a particular spatial distribution of vulnerability in contemporary Paris: a vertical geography of economic inequality that is matched by an unequal burden of risk. While many scholars have engaged with the biopolitical dimensions of urban form—that is, the ways in which the city, as a technological system, organizes and systematizes life toward the ends of a deployment of power by state and capital—few have explored the ways in which the vertical dimensions of urban space operate toward the differential valuation of life.[39] It is a valuation that "makes live and lets die," according to Foucault's description of the operations of a new form of modern sovereignty that first emerged in the late eighteenth century with the decline of absolutism.[40] While the structures of accessibility, comfort, convenience, and transport that date to the Haussmann period and the Belle Epoque encourage the flow of capital and community for those at the political and cultural center, a market-based violence pushes the already marginal—the elderly, the undocumented, the addicted, and the desperately poor—into the invisible spaces of the *chambres de bonne*.

When investigating the deaths of the heat wave's "forgotten" victims—those whose bodies remained unclaimed by family or friends, who lived and died alone in the city—I visited addresses distributed throughout Paris. The APSP maps showed clear concentrations of vulnerability in the southern and eastern districts of the city, and I had anticipated that my subset would largely follow this distribution. Yet although there were some parallels between the aggregate patterns of excess mortality shown by the APSP and the distribution of this group of the forgotten, a new spatial pattern emerged that appears to mark risk not necessarily for outright mortality, but rather for dying alone in the postindustrial city.

City officials collected the bodies of the forgotten from all but one arrondissement, from neighborhoods rich and poor, close to the river and atop Montmartre, in the city center and near the Boulevard Périphérique. The bulk of them were concentrated in administrative units with lower household incomes. The eighteenth leads the map with ten, followed by strong concentrations in the tenth, eleventh, twelfth, thirteenth, and nineteenth arrondissements. The surprises here are high concentrations of mortality in the fifteenth and seventeenth arrondissements. The fifteenth, while not especially rich, is a primarily middle-class district, while the seventeenth rivals the sixteenth for some of Paris's most elite addresses. Yet the seventeenth is an arrondissement divided, which also includes some of the lowest-income districts in the city, especially near the Porte de Clichy. And adjusting for population reduces the fifteenth's share of the burden significantly: with 234,091 inhabitants as of 2008, the fifteenth is by far the city's most populous arrondissement.[41]

The distribution of these victims by administrative unit tells only part of the story of a geography of vulnerability. Ten of the victims died in the eighteenth, the city's second poorest unit, but three died in the sixteenth, the city's richest;[42] adjusting for population, the mortality gulf between the two districts is not nearly as wide as the socioeconomic one. Far more important is a concentration of mortality on a different spatial register. Six in ten of the forgotten lived atop their neighbors in *chambres de bonne* or other marginal housing (including the four homeless victims). They include owners of these rooms, like Pedro, and renters, like Paulette.

Several critical factors unite these *chambres* as sites of vulnerability. The first and most obvious is the increased heat load of units on the upper stories of buildings. The *chambres de bonne* thus ironically serve as a protective buffer for apartments below: they operate as an insulating zone for the rest of the building in both winter and summer. Smaller apartments with only one window—at times a skylight rather than a wall-mounted window—suffer especially, with no capacity for cross-ventilation. The absence of baths or showers in most of these rooms leaves no easy means to cool off. And little social interaction means that there is small chance of anyone discovering a problem until it is too late. The most comprehensive epidemiological survey that followed the heat wave to date cited such conditions as highly determinant of risk.[43] In 2004 the InVS, a French epidemiological surveillance unit that is the rough equivalent of the U.S. Centers for Disease Control and Prevention, conducted a case-control study. Selecting some 250 victims and 250 survivors of the heat wave, the InVS carefully analyzed a range of environmental risk factors for death by heat stroke and dehydration. Principal among these were the characteristics of the victims' lodgings. Those who lived on the top floors of apartment buildings were four times likelier to die during the heat wave than their neighbors on lower floors. Those who lived in apartments with no bath or shower were two and a half times likelier to die than their neighbors. And those who had little or no social interaction were six times likelier to die than their neighbors. These social risk factors were as high or higher than serious medical ones, such as heart disease or high blood pressure.[44]

Yet just as important as these risk factors are those that link such properties to poverty and marginalization. The same report cited higher numbers of rooms per lodging and increased square footage as protective factors. The critical factor was not necessarily that apartments were located on top stories so much as that they were *small* apartments on top stories. The legal status of such rooms in the French building code indicates just how small such apartments can be. The code stipulates that for new construction, an apartment must have an area of at least 14 square meters and a volume of at least 33 cubic meters (at least 150 square feet, with a ceiling height of at least 7 feet 9 inches).[45] But the same code "grandfathers" in existing construction, and allows for the continued rental of much smaller rooms. These minimum requirements are for rooms of no less than 9 square meters (97 square feet) and a ceiling height of 2.2 meters (7 feet 2 inches). Apartments of more than one room are required to have a toilet and a shower or bath with hot and cold running water. But one-room apartments are exempt from the requirement of a bath or shower and toilet, required only to have access to a toilet in the same building—not necessarily on the same floor—and with no requirement for access to bathing facilities. Many such rooms, operating outside the law, do not provide running water in the apartment itself.[46] These factors have made the *chambres de bonne* the least expensive form of lodging in Paris, and given the city's desperate housing shortage, have made them the principal housing for the city's poorest residents.[47] Newspapers teem with stories about stiff competition among students and immigrant families for these rooms, often alongside other accounts of squalid conditions in the same apartments.[48] They offer reasonably priced shelter for those who can withstand the long climb up six or seven flights of stairs and a life with no amenities.[49]

Visiting these apartments provides critical insight into the challenges such dwellings posed for those who lived and died there. Their floors were lined with loose wires and leaky plumbing, exposed to weather through open skylights, and often teeming with pigeons and mice. For the young and able-bodied students who lived on the same hallways as Pedro in the Avenue de l'Observatoire, or Andre in the Avenue d'Italie, whom I met and interviewed in the course of my research, these conditions are inconveniences.[50] But according to 2006 census data, 10 percent of those who live in one-room apartments—many if not most of which are situated on top floors in older buildings—are over age sixty-five. For the elderly and disabled, a descent down a narrow spiral staircase is literally life-threatening. For Paulette, the eighty-year-old who lived in the Avenue de Friedland, retrieving mail a story below her rooms meant navigating a steep, narrow staircase; bathing meant carrying a heavy bucket from the common sink in her mouse-infested hallway (see fig. 19). Her final years were an assault on her dignity and a constant threat to her life.

Such conditions are not new. As a public health report from the early 1960s noted, the elderly increasingly found themselves relegated to such spaces as a function of their poverty. The reporter observed that "nearly all of the old live in buildings without elevators." Worse, his survey showed that "by some apparent irony, the occupants of the ground floor are proportionally more represented by the most able-bodied groups, and less by those whose physical activity is quite diminished." An intersection of debility and poverty meant that "the progressive weakness of many of the old will contribute to relegating them bit by bit into lodgings that are the least prepared to receive them."[51] A decade later, Simone de Beauvoir detailed what she called "a typical case" in *La vieillesse,* her indictment of the capitalist world's dehumanization of the elderly. Madame R., Beauvoir tells us, was seventy-five and lived in a fifth-floor, cold-water walk-up in the Marais. The water tap was "in a nook at the back at the top of a high step; it requires acrobatics, when one is not agile, to climb back down with a bucket." The toilets were "at the other end of the building." Madame R. had to "go down one floor, climb up another, and then climb up fifteen further steep steps. 'It's my nightmare,' says Madame R. 'Sometimes . . . I ask myself if I'll ever get back down'."[52] In the 1980s and 1990s, many elderly Parisians complained in demographic surveys about using bathrooms on different floors of their buildings and how the difficulties of the staircase led to their increasing reclusion in their apartments.[53]

For many of the forgotten, debility linked to age, illness, or both exacerbated the vulnerability with which poverty had confronted them. Economic desperation pushed them into marginal housing that combined with their

FIGURE 19. Sink in attic hallway, Avenue de Friedland. Photo by the author.

poor health to elevate their risk for heat death. Patricia's behavior betrayed signs of mental illness—also a significant risk factor for death during the heat wave.[54] At thirty-nine she had abandoned her family and fled to Paris without a word. In her room at the dingy residential hotel, she showed signs of hoarding, and her personal belongings indicated a religious fervor that bordered on fanaticism: she collected Bibles and other religious texts, and scrawled religious verse repeatedly in her notebooks.[55] Louise had aged in place like many in her cohort. The sixth-floor walk-up in the Rue Ambroise Thomas had been affordable when she moved in in her late twenties; it was all that she could afford when she remained there in her late eighties, no longer able to climb the staircase. She was surrounded by health care: she shared her building with doctors' offices and a pharmacy. But as with Paulette, economic deprivation kept her lodged in a scorching apartment under the roof, and increasing disability put the city and its resources out of reach. Pedro was perhaps the most active of the group. At age seventy-nine, he still attended Mass occasionally, and in 2003 he had attended a social event in his building's courtyard. But the others there said that he ate or drank nothing at the party, and talked to almost no one before returning upstairs. Since his wife's death a decade earlier, he had fewer and fewer contacts, and spent more and more time sequestered in his seventh-floor rooms.[56]

There are dozens of other cases like these: Robert, a fifty-six-year-old alcoholic who weighed some 260 pounds, climbing seven flights of stairs to his apartment in the Rue de l'Isly until his heart gave out on 11 August (fig. 20); or Sonia, an eighty-seven-year-old former domestic servant with a cold-water seventh-floor walk-up with skylights rather than windows in the Avenue Bugeaud, who died on 13 August; all lived in similar conditions and at similar risk. Each of these cases points to an intersection of poverty, disability, and the literally structural violence of an architecture that exacerbated the vulnerability of some of Paris's least resilient citizens during the disaster.

The association of the *chambres de bonne* with advanced marginality and health risk is long standing. Contemporary reports about poor families stuck between such rooms and homelessness are merely the latest echoes of similar cries for housing reform from the past century.[57] A report from the late 1950s calls the living conditions of families in the *chambres de bonne* "unacceptable," and epidemiological reports from the early twentieth century noted that the *chambres de bonne* were highly overrepresented in tuberculosis

FIGURE 20. Seven-story spiral staircase leading down from Robert's room, Rue de l'Isly. Photo by the author.

prevalence.[58] Moreover, the police have long projected the disreputability of such lodgings onto their residents. The *chambres* are a fixture of detective fiction and true crime stories,[59] and are now the sites of frequent police raids and identity checks: in the course of such raids in 2007, several children of undocumented immigrants without papers leapt from their apartment windows only to be critically injured.[60]

The *chambres* are thus a key site in a long history of poverty, marginalization, and disenfranchisement in contemporary Paris. They are an artifact of deep economic inequalities in the city, reflected in powerful health inequalities. It is all too easy to overlook the political economy of housing in emphasizing the physical dimensions of risk these dwellings pose to their residents. Contemporary discussions about the heat load on upper stories echo studies of tuberculosis a century ago, which attributed higher rates of the disease among residents of *chambres de bonne* not to the poverty that forced their residents to share insufficient space, but to ineffective ventilation. Such accounts are reminiscent of those that emphasize the physiological vulnerability of the elderly to heat stroke as the principal factor that explains why so many of the dead were over age seventy-five. This is clearly part of the story of aging and risk, but to accept these explanations in place of social and economic ones is to ignore a larger social truth. To attribute the vulnerability of those living in substandard conditions to increased heat load is to ignore the very real risks that attend economic marginalization, in particular when that marginalization is compounded by physical or mental disability.

City officials repeatedly decry the situations of families with no other recourse, but they have done little to mitigate the worst abuses. Despite wide agreement that the *chambres de bonne* are an unacceptable solution to the housing problem, to remove them from the market would deprive the city of some 20,000 housing units, exacerbating the existing housing crisis. Aside from placing families in public housing units often located in unstable neighborhoods outside the city limits, the housing ministry has effectively ignored the plight of those in such units. Yet the *chambres* remain an important space of risk and vulnerability in the urban environment, particularly for the elderly. They are sites of marginalization: repositories for those who have lived out of productive and social citizenship. In ordinary periods, they are miserable shelters tucked out of sight where the poor of Paris live alone. During the heat wave, they were sites, indeed agents, of a phenomenon of dying alone.

Epidemiologists and geographers have performed a critical service in mapping aggregate vulnerability in the city. These spatial distributions of mortality signal areas of general risk to the population during heat waves. Yet we would be remiss to focus exclusively on this macroscopic level, if only because it can blind us to the particular conditions that attended the deaths of the most desperate cases in the city. General distributions such as those the APSP produced reveal the relationships between space and the risk of dying during the heat wave, but the stories of the forgotten, and in particular, the stories of those in the *chambres de bonne*, point to the risk of dying alone. In turn, those stories point to the violence of everyday life that attends those living alone in such conditions day after day, year after year.

Conclusions

In recorded weather history, France has never experienced a heat wave of such intensity and duration. Although prior heat waves had brought excess mortality with them—notably in the summers of 1976 and 1983—France had never experienced anything like the 2003 catastrophe. The heat's arrival in conjunction with the August vacation period exacerbated the crisis, as many health care workers and state officials were away, inhibiting care and communication during the disaster. Likewise, the French population has aged since the 1980s, placing more people at physiological risk than ever before. Yet on another plane, the conditions of place—social and physical—that also intensified risk for some of France's most vulnerable populations have been in the making for at least a century. Life in the city's most desperate housing has long been linked with precarious health and marginality. While no one predicted the staggering mortality that the heat wave visited on France in 2003, the identity and geography of the likeliest victims—if not their number—were sadly predictable.

A range of social, economic, and cultural factors dramatically increased a broad physiological vulnerability during the crisis. But we must also recognize the ways in which the heat wave was also an ecological disaster that relied on a coupling of human and natural systems that is particular to the modern city. It was a period of meteorological extremity that exploited deep-seated vulnerability in the built environment. Disasters have specific forms of agency with different effects of place. Although they disproportionately affect those in the lowest socioeconomic ranks, floods, tornadoes, hurricanes, and heat waves exploit specific and different ecologies. Floods affect primarily those on the lowest ground; hurricanes those in coastal regions with insufficient building codes; and heat waves those who live on high floors.[61] As the environmental historians Ted Steinberg and Mike Davis have argued, vulnerability to contemporary disaster is socially and historically produced: place is a historical entity, produced through human decision-making and a nexus of geography and culture.[62] The poverty of the forgotten was the primary factor that structured their vulnerability, but it cannot be disentangled from the geographic factors that intensified their risk.

The geographer Susan Cutter has recently developed what she calls a Social Vulnerability Index. The index is a quantitative model that aggregates predictors of risk for those who live in areas marked by natural hazards: flood plains, hurricane zones, fault lines. Among the critical factors that shape vulnerability are age, poverty, disability, and poor housing quality.[63] Although her work has focused on the American southeast coastline, it has clear applications to other precarious geographies. The historical marginalization of the elderly and disabled poor to housing circumstances such as the *chambres de bonne* constitutes a distinct environmental risk, and merits attention as a matter of environmental justice. Environmental justice is in this view about all aspects of one's surroundings, including both human-made hazards to natural systems and the effects of community on physical, psychological, and social health.[64] Although frameworks for environmental justice in France remain weak and inchoate, the particular burden borne by the elderly and disabled poor as a function of place during the heat wave raises questions about the possibilities for forging a new form of environmental justice that envisions them as particularly interested stakeholders to be included in national and local decision-making.[65] Neither Haussmannization nor the *chambres de bonnes* produced the disaster. But the long-standing inequalities they both reflect and reinforce contributed significantly to shaping the heat wave's horrific outcome.

4

Vulnerability and the Political Imagination: Constructing Old Age in Postwar France

> Man never lives in a state of nature; in his old age, as at every other period of his life, his status is imposed upon him by the society to which he belongs.
>
> SIMONE DE BEAUVOIR, *The Coming of Age*

The Rue Cambrai sits in a depressed neighborhood in Paris's nineteenth arrondissement. There is almost no commercial activity aside from a small *tabac*, a pharmacy, and a café. Deteriorating housing projects from the 1970s constitute most of the cityscape. Evidence of vandalism abounds in the area surrounding the housing block where Muguette lived in August 2003. An eighty-four-year-old widow, she had lived in the apartment for decades, according to the *gardienne,* who spoke with me about Muguette in June 2009. The *gardienne* had herself lived there for more than twenty years, and knew Muguette well. As I exited the elevator to visit her apartment, I noticed that the number plaque indicating the floor had long since fallen off, and only a hand-scrawled "4" on the cinder-block wall told me where I was. It was here that Muguette died alone sometime in mid-August: although her official date of death is 17 August, city officials noted that her death appeared to have taken place several days earlier.

Not far away in the tenth arrondissement, Georgette G. died within a few days of Muguette. At ninety-seven, Georgette was one of the oldest of those who died alone during the heat wave. The retired seamstress had never married, and lived alone in her two-room apartment in a mouse-ridden building in the Rue Hauteville for more than thirty years. Her only social contacts were her physician and a domestic aide, Léa Mianfouna, a Congolese woman in her mid-thirties who visited Georgette twice a week to do her grocery shopping. Mianfouna was on vacation when the colleague covering her shifts arrived at the apartment on 14 August to a horrible odor in the hallway; she soon discovered Georgette's body in the bathroom. Mianfouna herself learned of Georgette's death only upon returning from vacation to find Georgette's door sealed by the police. Georgette was not an easy client, Mianfouna told the press: she had begun hoarding, frequently forgot to pay her bills, and refused to wash or change her clothes. Her extreme isolation troubled Mianfouna, who learned only from a neighborhood pharmacist that Georgette's sister had lived just blocks away, and had herself recently died without Georgette's knowledge. The aide described Georgette as "the most isolated" of her clients. "'I often said to myself, 'What if this were my grandmother?' At home, we never let an elderly person finish his life alone. That's a difference between Africans and Europeans that shocks me greatly."[1]

I cite these cases because elements of these stories appeared in two very different novels that engaged the question of aging in contemporary France in the aftermath of the heat wave. The first is Anna Gavalda's *Ensemble, c'est tout* (Together Is Everything).[2] Gavalda presents the reader with a redemptive story about solidarity conquering loneliness among four central characters. Much of the plot revolves around one of them: Paulette, a woman in her eighties who has lived alone with her cats since her husband's death. The novel opens with her suffering a fall. Gavalda describes how she has fallen before; for Paulette the falls are becoming more frequent than ever, and she has taken to hiding her bruises from inquisitive eyes. But this time a neighbor brings her to a doctor, who insists that it is time she go to a nursing home. She enters the home and a world of social misery against her will. Her grandson, Franck, finds visiting her unbearable: the moment he arrives he begins planning how to leave. But when he brings his girlfriend, Camille, on one of his visits, she develops a real affection for Paulette. Soon she insists they bring Paulette to live with them in Paris. Camille quits her job to become Paulette's full-time nurse. Camille takes her for walks, feeds her, and bathes her. When Paulette is shy about her aging body in the bath, Camille herself disrobes as if to show her own body's flaws. An artist, Camille does a series of studies of Paulette through the winter. When the next summer arrives, they bring Paulette back to her modest house in the country, where Camille stays with her. One summer evening shortly after the move, Paulette dies at peace in her garden.

The novel makes several clear references to the heat wave. The book's chronology suggests that the opening chapters are set during August 2003, and Franck warns Paulette against the stifling heat.[3] Later, an art dealer drops in on Camille and discovers her sketches of Paulette. He says: "These are great! Really wonderful! With the heat wave last summer, the elderly are really trendy now, you know? These will sell, I know it." Camille is "crushed" by his callousness.[4] But perhaps most important is the redemptive strain of Gavalda's narrative, which presents an alternative outcome to the horrific deaths in isolation that came to represent the heat wave in the popular imagination. Paulette suffers from the debilitation of old age and abandonment, only to find community and a peaceful death at the novel's end. The novel's dedication shows Gavalda's hand: "Muguette (1919–2003), body unclaimed." Muguette refers to one of the forgotten to whom Gavalda's mother, the visual artist Constance Fulda, had dedicated a canvas in her 2003 series of tableaux (see chapter 2). Muguette's life in a decaying project was a far cry from Paulette's. But the novel takes a narrative of solitude and brings to it an ideal of solidarity, whereby a group of characters—family and friends—take on the responsibility of Paulette's care and allow her life a graceful end.

The second novel appeared a year later, but is worlds apart in its representation of aging. The misanthropic novelist Michel Houellebecq's *La possibilité d'une île* (The Possibility of an Island) takes as its central premise not aging, but eternal life. The novel's protagonist, Daniel, is an unhappy, racist, misogynistic stand-up comic and screenwriter. From the novel's outset, we see that Daniel understands aging as a process of invariable decay. It repels him, as does his own aging body. He abandons a woman he loves because she is growing older; he then takes up with a teenager who in turn abandons him as an aging fool. In the process, Daniel becomes fascinated by a cult dedicated to the preservation not only of life, but of youth, through a cycle of genetic replication and suicide.

Daniel describes the early twenty-first century in Europe as a moment that saw "the increasingly ugly, deteriorated bodies of the elderly" as "the object of unanimous disgust." During the heat wave, he observes, the press published articles with "photos that were akin to those of the concentration camps, relating the agony of old people packed into communal rooms, naked on their beds, with diapers, moaning all day long without anyone coming to rehydrate them." Here Houellebecq, drawing on coverage of the heat wave in the French national dailies *Le Monde* and *Libération*, echoes Léa Mianfouna, Georgette's domestic aide:

> "Scenes unworthy of a modern country," wrote the journalist, without noticing that they were the proof, in fact, that France was actually becoming a modern country, that only an authentically modern country was capable of treating old people as pure garbage, and that such a contempt for the elderly would have been inconceivable in Africa, or in traditional Asian countries.[5]

Other throwaway lines are even more devastating. He describes one character as having that "senile, almost unconscious selfishness of the old."[6] He writes that for people in the twenty-first century, adulthood meant "worries, work, responsibilities . . . while ceaselessly witnessing—impotent and shameful, slow at first, then more and more quickly—the irremediable degradation of their bodies . . . until they were no longer good for anything and were definitively thrown onto the scrap heap, cumbersome and useless."[7]

In dramatic contrast to Gavalda's saccharine approach, Houellebecq displays a raw brutality that is difficult for the reader to bear. But his depiction is as incisive as it is reprehensible. If in Daniel's view the elderly are less than fully alive, less than fully human, he argues that French society had effectively embraced the same view, casting the elderly as marked for death and making possible the terrible and anonymous endings to the lives of Muguette, Georgette, and countless others. In Daniel's unspeakable view, the heat wave had constituted a sort of euthanasia program for relieving if not the victims, then at least the nation, from the misery of this population: the misery of paying for it, artificially sustaining it, even looking at it. Houellebecq's representation is extreme, but it provides insight into a widespread reading of old age as a problem of modernity that increasingly characterized French political culture in the postwar era. Since the mid-twentieth century, many French politicians, demographers, sociologists, and the press had developed a reading of the elderly as a population at the limits of citizenship and a burden for an emerging postindustrial nation. The aging of France threatened the republic's integrity: it menaced its economic vitality, it sapped its physical strength, and it generated an unproductive—and increasingly feminine—demographic that promised to break the welfare state. This chapter explores the ways in which for many in France, the elderly had become a vexing economic, political, and social problem, one that the heat wave revealed with stark clarity.

The "Problem" of Aging between Economics and Population Science

Historians such as David Troyansky and Patrice Bourdelais have argued that the mid-eighteenth century witnessed the emergence of old age as a distinct part of the life cycle in France. Whereas the political violence of the seventeenth century fed preoccupations about death, the relative peace of the eighteenth century allowed for the emergence of a concern with growing old and its implications for society. As the population itself began to grow in the middle of the century, so too did the proportion of the elderly in France. The state's emerging interest in population initiated a concomitant interest in the demographic structure of the population. Somewhat arbitrarily fixing the beginning of old age at sixty, the royal government initiated rudimentary retirement insurance programs, encouraging those who could to prepare for the last stages of their lives. At the same time, literature, the arts, and political discourse increasingly represented old age in the form of retirement as a reward for a productive life rather than as merely a forestalling of death.[8]

But it was more than a century later that a significant concern with the aging of the population began to take shape at the level of state and society. Combined with national fears of degeneration, France's devastating loss in the Franco-Prussian War sparked a crisis for many who were anxious about what they perceived as a problem of depopulation.[9] The realization of a birth gap between France and Germany and the staggering death toll of the First World War gave these concerns a fresh impetus. Pronatalist policies informed much of French political discourse through the interwar period on both sides of the political spectrum, with differences of degree rather than kind informing talking points on right and left.[10] In the interwar period most of this conversation—and consequent policy—focused on low birth rates: the state launched subsidies for large families and initiated the tradition of Mother's Day, adding to prewar policies such as maternal leave (and employment guarantees for women who opted to take it). But after the German Occupation, some of the critics who had led the pronatalist charge between the wars took up the cause of aging, which they saw as a critical factor in the development of a national vulnerability.

Fernand Boverat was among the most important participants in this debate. A far-right-leaning political crusader, at the age of twenty-eight he became the secretary general of the Alliance Nationale pour l'Accroissement de la Population Française (National Alliance for the Growth of the French Population), itself founded in the aftermath of the Franco-Prussian War. He was an admirer of Fascist population policies in Germany and Italy, and he published widely on pronatalism. In the interwar period he drafted extensive recommendations for consideration by the French National Assembly, including draconian legislation on abortion and contraception, which the Vichy government implemented in large measure during the Occupation. Even before the Second World War, he already saw depopulation as a problem at both ends of the life cycle: low birth rates and the aging of the population were the source of national weakness. Boverat framed the issue as a tandem problem of aging and feminization: as the population grew older, women also became overrepresented. A 1938 publication included as its frontispiece a cartoon showing Marianne, the figure of the French Republic, in her dotage, accompanied by the caption, "France will always be France! But even if she becomes a little old lady?"[11]

For nationalist demographers and politicians in the postwar era, aging threatened the vitality of the population. As Alfred Sauvy, another pronatalist who became the founding director of the Institut National d'Études

Demographiques (INED; National Institute for Demographic Research) in 1945, argued, the population could "grow" or it could "grow old," but not both.[12] For nationalist demographers and politicians, by growing older and withdrawing from productive activity, the elderly became a drain on a society desperate for economic recovery from the war; as parents of too few children, they bore responsibility for the nation's depopulation. In the aftermath of the Occupation, the elderly had become an icon of national weakness. "The conditions of modern warfare," one demographer argued in 1948, demanded armies comprised of "the young, in full possession of their physical means."[13] As two of his colleagues concurred, an "invasion" by the elderly had preceded the invasion of the Germans; thus "the terrible failure of 1940, as much moral as it was material, should be linked in part to this sclerosis."[14]

Seizing on this rhetoric, Boverat became one of the principal contributors to the debate on aging in France through the 1950s. For Boverat, the sheer magnitude of aging constituted France's major population problem. In 1949, Boverat wrote about his concerns to Georges Mauco, an extreme-right fellow traveler and sometime psychoanalyst who became the founding secretary general of France's Haut-Comité Consultatif de la Famille et de la Population (HCPF; High Commission on the Family and Population) in 1945.[15] The letter included what Boverat called a "small study on the increase in the number of very old," and indicated that "the Alliance nationale wishes greatly" that the HCPF would address these concerns.[16]

Encouraged by Mauco, Boverat prepared a substantial report for the HCPF in 1951. Titled "The Aging of the Population and Its Repercussions for Social Security," the report sounded the alarm about a national demographic emergency. Boverat noted that since 1775, the proportion of those under twenty had declined by 50 percent, while the proportion of those over sixty had increased by 228 percent. More troubling was the increase in the "very old." The past century had witnessed a 65 percent increase in the number of those in their sixties, but a 122 percent increase of those in their seventies and a whopping 200 percent increase in those in their eighties. As a consequence, "the French population is currently the oldest in the world." Particularly alarming for Boverat was that "the number of elderly women" was "much more considerable and has grown more than that of elderly men."[17] Since women traditionally spent fewer years as wage earners than men, they were therefore "a heavier burden for the collective."[18]

Boverat attributed the rise in the number of elderly citizens to advances in medicine. He argued that such progress was a good thing, "but the significant increase in the number of elderly should not be lost from view." Caring for the very old constituted an increasing burden for the nation that should dampen enthusiasm for longevity. "The growth of the number of elderly people would be a good thing if we could assure for them a satisfactory life," Boverat argued. Yet this assurance was far from certain, as the growth of the elderly as a proportion of the population came at the expense of the working-age population. As a consequence, the elderly became a parasitic population, "consum[ing] a share of national production without contributing to it."[19] Far from benign, aging threatened the country economically and socially. For Boverat, "by slowing economic activity, aging halts production and impoverishes the country."[20] Moreover, aging was a problem with cascading effects: aging slowed national rates of marriage and reproduction while increasing rates of mortality.

Among Boverat's principal concerns was the economic effect of the aging population. Indeed, the economic implications of an aging population have not only shaped contemporary views of the problem—particularly in light of France's development of a comprehensive social security system in the mid-twentieth century—but also informed the historiography of aging in modern France. Historians and sociologists who have studied aging in contemporary France have focused principally on the labor and economic histories of the phenomenon. They have read industrialization and emergent capitalism as the driving forces behind the modernization of aging. According to this literature, productivity and one's place in the economic life cycle became critical markers of value in an industrializing era. The mid-nineteenth century witnessed a dramatic scaling up of the division between family and business. Where the aging of workers had been a family responsibility in a preindustrial era—even for artisanal occupations—the industrial period's atomization of the family made this arrangement increasingly impracticable. For the sociologist Rémi Lenoir, the advent of the state's interest in managing the survival of the elderly at this point introduced the modern framing of old age in the economic terms of retirement.[21] By the end of the nineteenth century, this had become an acute question for syndicalists as increasing numbers of aging workers faced the problem of retirement in a moment that, for Marx, had succeeded in "throwing every member of [the] family onto the labor market."[22] Yet framing aging in the guise of retirement also linked old age inextricably to questions of labor and economic production, placing the elderly in a difficult position. If syndicalists, concerned for their own futures, became the primary advocates for state- and capital- sponsored retirement, they also experienced a powerful conflict of interest that set the protection and job security of elderly workers against the interests of younger workers in an era often marked by staggering unemployment.[23]

For the first time the state now had to grapple with the problem of a growing population of those who had aged out of their capacity to work. In the mid-nineteenth century the state created incentives for workers to save earnings in preparation for the arrival of the disability that would likely attend their old age. But workers objected to legislation that proposed mandatory pay deductions for retirement savings: work itself was so precarious, and wages so low, that the sacrifice of immediate economic resources for the sake of an abstract retirement far in the future was almost unimaginable. As a number of historians have argued, retirement was low on many labor activists' priority lists, far below issues such as job security, a living wage, and full employment—because few workers expected to live long enough to enjoy retirement.[24] Given the still low life expectancy in France even in the early twentieth century—fewer than 10 percent reached age sixty-five—union leaders saw proposed pensions that fixed the retirement age between sixty and sixty-five as more like life insurance plans to provide for their families after their deaths than real retirement plans.[25] The Socialist deputy from the mining region of the Nord, Jules Guesde, argued that a retirement age of sixty-five "is, effectively, retirement for the dead, at least in some industries, where no laborer reaches such an old age."[26]

Yet retirement still occupied a place on the political agenda. As the historical sociologist Anne-Marie Guillemard has argued, union leaders sought to reframe the debate over retirement from the idea of insurance against disability—which would require laborers to pay into savings accounts—to a just reward for a life of labor. Just as workers sought to limit the length of the workday and the workweek, so they sought to place limits on the length of a working life. They were rewarded with piecemeal legislation beginning in the late nineteenth century. In 1884, for example, workers in the mining industry won the right to a state-backed pension, with railway workers and state functionaries soon to follow.[27] The first universal benefit for the elderly was, by contrast, not a pension, but a welfare bill that provided for the elderly, the infirm, and the incurable, which passed in July 1905. Yet this bill, in contrast to a pension, was predicated on incapacitation. It defined old age as a form of disability rather than a life stage that all might hope to attain. It also offered merely a pittance to its recipients, with a threshold for eligibility. The law was particularly hard on widows, requiring not only the widow but also her late husband to have achieved a certain age and to have worked a certain number of years, and barring any benefit in cases of divorce, separation, or the survival of a child with an income, regardless of whether that child agreed to support the widow.[28]

A proposed pension law in 1910 sought to remedy the shortcomings of the 1905 law, but neither workers nor industrialists supported the bill, each seeing it as too much of a compromise. It was only after the Second World War that a real social security system emerged in France, as it did throughout Europe. But the rise of the welfare state was more than merely a public investment in the population's well-being and economic and health security, as it was in Britain or the Scandinavian countries. As the historian Paul Dutton argues, the French social security system was cobbled together from both preexisting and new public and private insurance and taxation plans, originated in large part by employers and distributed unevenly according to industry. A long history of industry-labor relations and the special role played by mutual aid societies, or private insurance cooperatives that came to occupy some official role and state sanction but remained primarily private entities, gave shape to the French social security system, which continued to prioritize repopulation and family assistance over retirement.[29]

For good reason, then, much of the historiography has focused on the connections among labor, economic security, and old age. It has constituted primarily a labor history of retirement rather than a history of aging per se. But although the economic framing of aging through the question of retirement is critical, it leaves important cultural and political dimensions of aging unquestioned. For Boverat and others who worried about aging in postwar France, the economic dimensions of old age were paramount, but did not constitute the entire discussion of the problem. The debates over pensions and a right to retirement also framed the representation of aging in specific ways. If the law of 1905—the first widespread assistance to the elderly—was predicated on the inherent disability of the aged, it reinforced the notion that the elderly were a population apart. And they were so permanently: the law irrevocably bound old age to infirmity and incurability.

As early as the seventeenth century, the French distinguished between a "green" or "vital" (*verde*) old age and a "decrepit age."[30] Although this distinction faded away in the eighteenth century, the retirement debate brought it back in full force. As Elise Feller argues, a struggle ensued through the early twentieth century to recast aging as a legitimate and fully vested form of life. The securing of the right to a pension for most French workers successfully reframed the "old man" and "old woman" (*vieillard*) into the "retiree," engendering a new category of citizen.[31] The sociologist Vincent Caradec notes that this reclassification is the result of a "movement toward the *institutionalization of the life course*," through which the stages of life are formalized with increasing specificity through the individual's relationship to the state in modern society.[32] Yet as Caradec and others have also indicated, such moves are merely new modifications of a long-standing distinction between the capable and dependent forms of old age. If the "Third Age" marked by the beginnings of retirement has positive connotations of activity, leisure, and enjoyment, the "Fourth Age" that follows is one of dependency and debilitation.[33]

This distinction between capability and dependency was at the heart of the concern over aging that marked France in the mid-twentieth century. Yet it merely finessed the larger problem of population in mid-twentieth-century France. For even if a new category of "retiree" had begun to replace that of the dependent elderly, it still marked a way station between working adulthood and what critics came to call true old age. In Boverat's and others' view, only the state was equipped to remedy this problem, through prophylactic action in order to minimize its disruption to the nation. Boverat, and subsequently a national commission, sought to calculate the ways in which the state and capital could manipulate the national body in order to immunize it against the dangers of shifting demographics. For Boverat and other critics who feared the aging of the French population, nothing less than national vitality was at stake. As an inevitable biological phenomenon, the aging of the population threatened to sap national economic and political strength, calling for the deployment of an arsenal of legislation designed to manipulate the state of the population and restore national vigor.

The philosopher Michel Foucault's introduction to the concept of biopower as a technique of modern sovereignty provides a useful frame for interpreting Boverat's language and the strategies he advocated. For Boverat, aging was a natural process, but one subject to the influence of the state. This capacity of the state to direct life, as Foucault argued in his lectures at the Collège de France in the late 1970s, contrasted with the lethal power of the sovereign that characterized the early modern period—whereby the state, in the figure of the monarch, had the power to kill those who threatened sovereignty. In a new formulation, power instead consisted in an entire apparatus that could shape both individual lives and the general structure of populations. This new power, which emerged in the late eighteenth and early nineteenth centuries as a function of the political modernization of the state, operated at several levels. At the level of the individual body, disciplinary institutions such as schools, prisons, and hospitals shaped behavior and molded individuals into a coherent whole. The biopolitics of populations, as a new "technology of power" more closely linked to the state, operated at the aggregate level. The emerging notion of "security," as applied to populations, was looser in its operations than the discipline of individuals: it aimed at massaging or manipulating the edges of enormous problems, gradually influencing means rather than enforcing norms.[34] Moreover, it operated explicitly within a context of liberal democracy. Population did not respond to decree: it was instead a "thick natural phenomenon," something that could be adjusted rather than coerced, varied, or confined by the agency of the state and capital: "Population varies with the climate. It varies with the material surroundings. It varies with the intensity of commerce and activity in the circulation of wealth. Obviously it varies according to the laws to which it is subjected, like tax or marriage laws for example." Where disciplinary institutions aimed at the firm rehabilitation of individuals—the sick, the vagrant, the barren—the biopolitics of populations sought to gradually influence levels of sickness, poverty, and reproduction in themselves. A subtlety of action marks the biopolitical. Population is "accessible to agents and techniques of transformation, on condition that these agents and techniques are at once enlightened, reflected, analytical, calculated, and calculating."[35] Biopolitics thus acts at the level of statistics: rates of mortality and morbidity, yields of internment and rehabilitation, a management of risk, and a maximization of reward are its principal registers.

One lecture in 1976 explicitly engaged the question of a biopolitics of aging. Here, Foucault echoed Marxist readings of capital's mediation of the family, noting that a technique of government through the deployment of discourses of public health, hygiene, and population first seized on aging as an important site of intervention as a function of the rise of capitalism. For Foucault, the rise of the industrial sector initiated a deep concern for the ineluctability of the process of aging and its inevitable extraction of subjects from "the field of capacity, of activity."[36] But this transition was not merely economic. Illness, death, and aging were "phenomena affecting a population." Like illness and death, aging was "something permanent, something that slips into life, perpetually gnaws at it, diminishes it and weakens it," a problem that "can never be eradicated," that acts to "incapacitate individuals," and that demands the development of new techniques of management: "subtle, more rational mechanisms" of "insurance, individual and collective savings, safety measures, and so on."[37]

Aging was thus a biological phenomenon that could not be eradicated. It had a "thick naturality" that marked the population through its permanence. But that naturality was still subject to state intervention and manipulation on several levels. The state's institutions had the capacity to direct life and so influence the course of national demographic trends. As Boverat argued in his report, "It is not possible to modify chronological age and prevent a man who has lived 64 years from being in his 65th." Instead, Boverat advocated exactly the sort of policies that Foucault highlighted decades later as a means of influencing the population's center of gravity back to a more youthful median. "The recruitment of young foreigners of the white race" represented one possibility, although a limited one. Another was to ensure that the state did not "increase the difficulties of the problem to solve: that is what we are doing by prolonging education, lowering the retirement age, and reducing the number of work hours": each of these moves kept the able-bodied out of the active workforce. But the most important mechanism the state and its institutional allies could adopt was the capacity to shape life constructively in order to combat the pathogenic forces that aged the population prematurely in a physiological sense. Boverat argued that "certain significant factors in premature aging can be eliminated, the influence of many others can be reduced." Living a healthy lifestyle was critical, as alcohol and tobacco abuse wore on the body, advancing the ravages of time. Likewise excessive fatigue contributed to premature aging. Effective self-discipline and eschewing bad habits were important at the individual level for the worker, but "what he can do for himself to delay his aging must be completed by the action of public powers, and by that of his employers if he is in the private sector." Meaningful shaping of the population demanded the alliance of state and capital to this end. Extensive alcohol prevention policies, tuberculosis and other infectious disease control, labor protections, and housing policies could all combine to inhibit the dangers that aging presented to the population.[38]

Boverat supplemented his report with direct policy recommendations that he submitted to the HCPF in the form of draft legislation. "In order to halt the statistical aging of the French population," the motion asserted, it was essential to develop family and housing legislation that would encourage young families to reproduce "in order to divert the new rapid drop in natality, a drop that, if it continues, will accelerate the aging of the population." The legislation also recommended "that the struggle against infant and child mortality be intensified" and "that a select immigration of young foreign workers, easily assimilable and by preference the heads of households, be organized to fill the gaps in the active population—who support the care of the old." Finally, the draft sought "to improve the health and aptitude for labor for workers who are advancing in age" through a "struggle against alcoholism" and "fatigue."[39]

The direction of life represented one side of biopower; the other involved allowing death. As Foucault argued in *The History of Sexuality*, the transformation of modern sovereignty entailed a shift from "the right to *take* life or *let* live" to "a power to *foster* life or *disallow* it to the point of death."[40] The manipulation of population involved both of these possibilities: the fostering of youth, adulthood, and vigor, and the disallowance of life *at* the point of death, life that had reached the limits of its vitality. Here, Foucault emphasizes the importance of killing in the name of life. Political discourse in the modern period has often focused on the importance of killing—through the mechanism of warfare—not so much in the name of the conquest of territory but rather for the preservation of life. Regimes dehumanize others in the name of saving the lives of their own, feeding life by stoking death. But there is another, subtler side of what some have called a "negative biopolitics."[41] In directing resources toward the lives of some, the state effectively cuts off the lives—and social citizenship—of others.

In 1895, the pronatalist demographer and statistician Jacques Bertillon implied that a generational struggle had made France's birth rate the lowest in Europe. For Bertillon the reduction in mortality that France had experienced since the mid-nineteenth century was not cause for celebration, as it had only contributed to the aging of the population. Rather than something to be mourned, death was itself necessary to new life: only the death of the old could produce the life of the young. "No matter what the age of a death," Bertillon argued, "we can explain easily how the death provokes a new birth":

> Is it a child? His parents will experience the need to direct their affection toward a new life. . . . Is it an adult? The children that he could have had are already born, and, for another thing, his place in the sun is now occupied by another, younger person, who takes advantage of the occasion to marry and, in turn, to have his children. Is it an old man? If he is poor, he constitutes a burden for his family, who will be relieved and his death will allow more easily for the education of another child. If he is rich, he leaves an inheritance that allows his heirs to marry and have children. Thus, every death, at no matter what age, leaves a vacant space, a place quickly occupied by new births.[42]

Bertillon specifically engaged the ways in which an aging population in particular inhibited the growth of the population, offering the analogy between forest management and policy. "When the lumberjack makes clearings in the forest, the trunks and shoots bud everywhere, and the forest reconstitutes itself, without anyone having to take care of its growth," he argued. "But what to say about someone who . . . imagined nothing but eliminating the lumberjack's hatchet and saving the trees indefinitely! He would only wind up uselessly aging his grove, and in the end, would be beaten by this fight against death; because the law of living societies, for forests as well as for nations, is the perpetual renewal of beings." Bertillon insisted that the physicians and hygienists who sought to improve health and prolong life resembled the latter "ignorant forester." They were to blame for France's population problem, because "aside from the fact that it can only reduce mortality by an insignificant amount, this same reduction in mortality could only have the effect of diminishing further the number of births."[43]

A half century later, Boverat (who, like Bertillon, worked within both a pronatalist and a eugenicist paradigm) demonstrated a greater subtlety but ultimately the same message. His 1951 report flirts with several disturbing possibilities, moving gradually toward a discarding of the humanity of the aged. Reflecting on the disproportionate increase of the elderly in France's population, he insisted that "only a significant increase of young foreigners or a massacre of the old could perceptibly reduce this growth."[44] He went on to note that increasing immigration was only a temporary measure; in the meantime, "the care of the old will become more and more onerous" with medical progress, bringing higher expenses. At the beginning of the report, he reluctantly celebrated the human decency that would prevent a wholesale consignment of the elderly poor to death: "It is unlikely (thankfully) that we will resign ourselves to letting the old, for whom it might be possible to prolong their lives, die for want of appropriate treatment."[45] But such humanism was a costly indulgence. "At least if we do not want to leave a very considerable number of the old in the street in 10 or 15 years, or if we don't want to condemn them to finish their lives lamentably in the slums, we will have to build or refit retirement homes for hundreds of thousands of them." In addition to the staggering cost of such a proposition, he asked, how could the country staff these facilities, given the decline of the working-age population? By the end of the discussion, he reframed the question of insuring the health—and lives—of the aged: "Could we," he asked, "refuse to give the old treatments that are capable of prolonging their life for several years?"[46] What was "unlikely (thankfully)" at the report's beginning has now become a rhetorical possibility.

Boverat was one actor among many who recognized aging as a primary threat to the republic. Although Mauco and the HCPF did not act on his proposed legislation, they did establish a study group to examine the question of aging, directed by Pierre Laroque. Laroque was a specialist in social welfare, having served on a social insurance board in the French cabinet in the early 1930s and as the director general of the French social security division since 1944. In 1960 Mauco named him director of the Commission d'Étude des Problèmes de la Vieillesse (Commission for the Study of the Problems of Old age; henceforth the "Laroque Commission"). The commission's activities reflect a growing awareness that aging had undergone a major transformation and was now an issue that commanded national attention. Its work marks a shift from the notion of national decadence and old age—the sentiment that the elderly weakened national vitality—to the idea that they signaled an economic and political danger to the Fifth Republic at a dynamic moment in its development, what historians have called the "thirty glorious years" (*les trentes glorieuses*) of economic development.

The Laroque report, submitted to the HCPF in 1962 and published widely, is ambivalent about aging. It vacillates between a sympathetic analysis of the real difficulties faced by the elderly and the possibilities of the state to manage those difficulties, and a language of objectifying difference portraying the elderly as a drain on social resources and a stultifying force. From its outset, it betrayed a significant bias: the very name of the commission conceived of aging as a "problem" for the nation, rather than a "question" or an "issue." The report concerns itself with the demographic reality of an aging population and its meanings for the welfare state: its principal focuses are issues such as the retirement age and pensions, housing for the elderly, and social assistance. Its opening lines signal that "the aging of the population entails consequences in all domains of national life; progressively, but in an unavoidable way, it is burdening the conditions of existence for the French collectivity."[47] Aging also menaced the dynamism of economic and cultural development, as "politically and psychologically, aging means conservatism, attachment to old habits, the failure of mobility, and an inability to adapt to the evolution of the current world."[48]

The major issue was not aging in its own right, but the extent of demographic change. Although aging had been on the rise incrementally, it had now achieved such a scale as to transform French culture and society. The report indicates that the population over sixty-five had doubled between 1851 and 1954, but the total population had increased by only 20 percent. In other words, "it is striking to note that if, in a century, the total population has accumulated nearly 3 million old people, it has only gained 53,000 children."[49] It was thus the magnitude of aging that demanded a state response. In the "patriarchal societies" of the past, the report argued, communities and families managed the small number of elderly with which they contended. But in the modern world, "such a state of things has become exceptional." Now, and particularly "in industrial and urban environments, the conditions of professional life, of housing, the disruption or, at least, the breaking down of family connections between successive generations has made it necessary to seek solutions adapted to the specific needs of the elderly, who can no longer find their satisfaction in a traditional life framework." Where previously family and village cared for the elderly, "the intervention of public authorities in the domain of aging has become, in these conditions, indispensable."[50]

The bulk of the report tried to skirt a fundamental ambivalence of inclusion and difference in its approach to the elderly. It signaled the arrival of a new age that necessitated a state policy on aging, including guaranteed pensions for retirees and a massive expansion of state-subsidized housing for the elderly. The project represented a massive expansion of social citizenship that for the first time would include the elderly as a specific target of state intervention. Yet in so doing, it also designated the elderly as a group apart. The elderly became in this formulation what Adriana Petryna, writing in a different context, has described as "biological citizens," or those whose primary link to the state results as a function of their poor health or other incapacity.[51] The report—and the state programs that ensued in the following decades—fell into a double bind: efforts at inclusion created a fundamental segregation.[52] The commission had sought from the outset to avoid this trap, indicating that it aimed "at an adaptation without segregation": a policy that would provide for the elderly without rendering them a population apart. The commission wrote that "it is even preferable to renounce a too-perfect adaptation if it can only be achieved at the cost of isolating the elderly." An aging policy, it insisted, had to be "only one aspect of a larger policy that aims at assuring a harmonious development of all of society, with a view toward allowing each to occupy, at any time, the place that assures the most complete flourishing of his personality, in his own interest as in the interest of the community itself, accounting as much for age as for other elements that determine this personality."[53]

Such language suggests the commission's adherence to an optimistic imagination of human development that characterized the postwar era, with roots in the United Nations Charter, the Universal Declaration of Human Rights, and the Constitution of the World Health Organization.[54] But combined with the documentation produced in the commission's work on the report, it indicates a painful awareness of this dividing practice in action: that society was remaking itself through the marginalization of the elderly, pushing a growing population to the limits of citizenship. To produce the report, the commission charged its members with the production of extensive data about a range of problems that concerned the elderly, including the financing of retirement, the state of housing for the elderly, and the physical and psychological effects of growing old. Commission members interviewed dozens of stakeholders, including leadership of France's trade unions, social services providers, charity groups, and national ministries that provided for social welfare and economic development. The commission also engaged the services of the Institut Français d'Opinion Publique (IFOP; the French Institute on Public Opinion), a polling agency modeled on the American Gallup organization, to glean representations of aging from a survey of the wider population. IFOP polled a nationwide sample of 3,000 citizens, each of whom completed an extensive questionnaire with nearly eighty questions on aging and retirement.

The results of this work were sobering. Regardless of what the Laroque Commission had set out to do, it was clear that the nation had made up its mind on aging. Far from imagining old age as a future stage of humanity—that all citizens were potentially elderly—the results of polls and reports showed that the bulk of the nation considered old age to be a state of alterity, and many elderly experienced aging as a process of increasing isolation and marginalization. Some saw in old age the benefits of experience and wisdom: for example, four of five surveyed thought that the elderly had important experience to share on the job, and three of four saw the elderly as in general more conscientious than the young. But over half of those surveyed thought that the elderly were both unable to keep up with the pace of the modern workplace and lacked the ability to adapt, and more than a third thought the old were keeping their jobs to the detriment of the young.[55] Many elderly expressed a desire to live with their adult children and their families upon retirement. But the younger groups surveyed thought this would work only in "strictly utilitarian" conditions—that is, if their aging parents could provide domestic service and babysitting for the family.[56] Perhaps the starkest indication of a segregative mentality among the young was a dissonance on the question of nursing homes: 72 percent of French adults saw the expansion of nursing homes and senior communities as "desirable," but only 22 percent imagined themselves living in one: they were too "brutal," too "sad," "demoralizing," and in them "the old are treated too much like animals."[57] From a few intellectuals' condemnations of aging and calls for the rejuvenation of France at the turn of the century, national opinion on the elderly had, at least rhetorically, begun to throw the elderly onto the "scrap heap" that Houellebecq described in 2005.

Market practices and public opinion contributed to this dehumanization of the elderly. One report by a social worker that formed part of the commission's deliberations noted that as the elderly aged, they became poorer and forced to make do with increasingly inadequate housing, as they lacked the financial resources to move. Most of them therefore lived in unsuitable conditions ("apartment on the 3rd/4th floor, even sometimes in the attics"; "toilet, sink on the landing, on a lower floor, even on the ground floor").[58] Such conditions were normal enough, especially in working-class districts, as the Laroque report noted.[59] But as noted in chapter 3, these conditions were cruel to many aging bodies and facilitated the marginalization of the elderly by rendering them prisoners of their own homes. The demographer Alfred Sauvy noted wryly during the commission's deliberations that this phenomenon encouraged a segregative mentality that would push the elderly out of consciousness. This, he said, "has long been the norm for social programs for the elderly: get rid of the old—humanely of course—but cut them out of the loop because they get in the way."[60]

Perhaps most important is that this was not just a perception of the young. As the commission discovered in the course of its work, many elderly felt much the same way, having internalized a broader rhetoric of dehumanization. A report from the national social workers' union described many elderly as deeply anxious about "what they are becoming": the report described "isolation" as "the sentiment most often felt by the elderly person. When they are no longer supported by their family, nor by their intellectual life, they become reclusive."[61] In an effort to remain independent, they often refused help from their adult children, not wanting to be a "burden" to the family.[62] One seventy-eight-year-old woman whom social workers interviewed described her apartment as a "refuge, a shell that protects me from a hostile world." A man, eighty-one, suffered emotionally above all from isolation and dependency: not having had children, "not having any more family, nor any real friends, has become reclusive, hides his misery, and is humiliated at the thought of asking for anything."[63]

A critical question is why aging became such a preoccupation of the state in the postwar period. As the population had been aging incrementally for at least a century, why the sudden concern? Part of the answer is that for some, aging had long been a significant problem, but in the interwar period, with looming fears of a new war with Germany, fears of a declining birth rate trumped those of an aging population. The return of peace allowed for a focus on the other end of the life spectrum. Fernand Boverat remained concerned about both problems. His major criticism of the Laroque report was that it left natality underdeveloped as a solution to the problem of aging.[64] But the baby boom of the late 1940s and 1950s alleviated most concerns about the birth rate, as did a dramatic reduction in infant mortality.[65] Aging thus appeared as a primary threat. Moreover, with the development of a national social security system in the 1940s—which witnessed a dramatic expansion of the welfare state, including the first truly population-wide pension plan—a problem of the family or the market now became a public responsibility.[66] Finally the sheer magnitude of aging made it an issue the state could no longer ignore.

The more complex question is not only why a state policy for aging emerged in the 1960s, but why a sort of negative biopolitics emerged alongside it. That is, it is relatively clear why the state attempted to manage its population imbalances by fostering life; but how did the aged begin to lose their humanity, to creep firmly toward the margins of citizenship and no longer count as part of the population in the social imagination—that is, toward Houellebecq's "scrap heap" rather than toward Gavalda's integration of the elderly into a multigenerational social fabric?

As a number of historians and sociologists of aging have noted, the denigration of the elderly was nothing new in this period. In Peter Stearns's words, French society has traditionally treated the elderly with "almost unmitigated disdain."[67] Yet there was a kind of renewed vigor to these representations that emerged alongside a comprehensive welfare state in the mid-twentieth century. At the moment that the "old" had become "retirees" (on their way to becoming "seniors"), a broad social contempt created a significant obstacle to the state's efforts at inclusion of the elderly.

The Laroque report signaled that this was in process in its concerns about the segregation of the elderly. It concluded by presaging Houellebecq's caricature of the press in *La possibilité d'une île*, noting that France had entered "a moment when the misery and the suffering of the aged population are no longer tolerable, no longer worthy of a civilized country."[68] Laroque's colleagues agreed: Sauvy, of course, who saw a national meme in the effort to "get rid of the old," but also Georges Mauco. In a 1963 article in the Swiss newspaper *La Liberté*, Mauco wrote: "The old suffer from a disconsideration that is heavy with consequences. The word 'old man' often evokes an image of unpleasant or pitiable senility."[69] The size of the elderly population was unprecedented, the state was now responsible, but what was perhaps most important was that the imagination of the elderly had changed in the postwar era. The daily *France-Soir* put the debate in stark terms that reflected a brutal reality of representation: "Do the elderly have the right to live?"[70]

One critical element of the notion of the elderly as a population at the edge of humanity was the idea of their unproductivity, their "non-utility" in the late industrial economy.[71] As the sociologist Robert Castel has argued, the modern period has witnessed the rise of the *société salariale*, in which social citizenship has become affixed to status in the marketplace of labor. Rather than *exclusion* from citizenship—which he associates with a permanent state rather than a process—Castel describes a mechanism of *disaffiliation* from labor characterized by "passing from integration to vulnerability, or sliding from vulnerability to social nonexistence."[72] As such privileges of citizenship as health coverage, family assistance, and pensions became closely linked with one's place in the workforce, distance from wage labor through disability, immigration status, or age marked a degree of segregation from full social investment. Retirement—a life stage defined by the absence of work—framed the elderly as fully outside the labor market, and therefore pushed to the margins of a new society organized around economic productivity. In addition, in the transformation to a consumer economy that marked the *trentes glorieuses*, the elderly were nonconsumers in the market—a population on fixed incomes who "saw their savings disappear bit by bit"[73]—who disproportionately consumed public resources. Although there was a modicum of understanding that the elderly had aged beyond the capacity for self-sufficiency, Castel notes, a lingering market-based logic also saw the dependent elderly as responsible for their own lot, a result of insufficient saving during the period of work.[74]

Another critical element was the rise of a youth culture on whom the new consumer market depended and in whom the nation increasingly invested. The baby boom transformed France's demographic imagination at least at the level of representation, and a state commitment to family and youth marked this population as central to the national mission, brushing the elderly aside.[75] As the Laroque report acknowledged, "The aged population has been sacrificed by French social politics in the last fifteen years," which had aimed "at the encouragement of natality": France had been driven by "the necessity of favoring the young."[76] But the apotheosis of youth went far beyond the state's investment in the family.[77] As Mauco argued, it was "the development of values that valorized the child" that had "emancipated youth" and "forced the retreat of the notion of patriarchal authority and greatly reduced the prestige of adult authority, and of the elderly even more so." State policy and social practice had initiated an unprecedented generation gap in France, with youth now defining itself against old age and pushing it to the margins of humanity, as in the "contemptuous flippancy" of a new youth vernacular about the elderly as "shrunken, slumped, fogeys," a vernacular that "disvalued," "rejected," "segregated," and denied them their "human capital."[78]

Given the emphasis that the Laroque Commission and the social debate that ensued placed on the economic dimensions of aging, it is tempting to cast the issue in a frame that privileges the market and the alienation of the traditional family. Indeed, Rémi Lenoir and other social scientists have done exactly that, Lenoir arguing that the Laroque Commission merely "scaled up the problematic" of retirement and social welfare that had marked the modern industrial era.[79] Yet the commission's work and the debates within which it operated also engaged questions that went beyond economics: questions of vitality, humanity, and citizenship. The commission both questioned and reified the notion of old age as a limit of humanity, bearing witness to an increasing representation of aging as decay and devaluation of life. The question—the problem—of aging in this period therefore also merits a biopolitical framing that examines the conditioning of life at its limits and its links to processes of dehumanization, as well as the ways in which the state and other institutions both deploy and instrumentalize discourses of health and hygiene toward the end of encouraging, nurturing, and developing the lives of some and allowing others to exist outside of its frame: how they make live and let die.

Castel's notion of "disaffiliation" goes a long way toward connecting the economic components of aging and retirement to a broader politics of resentment that has pushed the elderly to the margins of citizenship since the early twentieth century. Yet the castigation of the elderly—and the cavalier attitude toward their disability and death that one finds again and again in the period—is not merely economic. It moves beyond Castel's category of "social nonexistence" and into the domain of what the Italian philosopher Giorgio Agamben calls "bare life," or life at what Michael Hardt and Antonio Negri have called "the negative limit of humanity."[80] Borrowing from Hannah Arendt, Agamben points to a Greek etymological particularity that distinguishes between at least two forms of life, *bios*, or the biographical life of the subject, and *zoë*, or the bare existence that is common to all living animals. Although Agamben writes of a number of forms of dehumanization under the law, including the plight of political refugees in the present, his primary concern is with specific political conditions under totalitarianism: a sort of politics of death that erases the vital legitimacy of certain populations on the grounds of their absence of humanity. He orients his focus around the rhetorical, legal, and epistemological process of dehumanization, a sort of reimagining of the figure of the biographical into the hollow specter of *zoë*.

Yet despite some important limitations, some of the implications of Agamben's work might apply to the epistemological violence that undergirds the notion of aging as a pathway toward an existence without life. If elderly French citizens began their lives fully politically vested, they have not ended their lives that way. Instead, the legibility of their citizenship faded in the course of their lives. The bestialization of the aged—"The old are treated too much like animals"—and a broader language that conceptualizes them in terms of annihilation—"segregation"; "Get rid of them"; "Could we let them die?"; "Do they have the right to live?"—signal some of the ways in which the notion of bare life is a useful one for imagining the trajectory of old age in the postwar era: a development that arrives, despite the commission's efforts to encourage a "flourishing" of humanity, in a rejection of its possibility.

"A corpse under a suspended sentence": Growing Old in the 1970s and 1980s

The Laroque Commission concluded with extensive policy recommendations, many of which the state implemented in the decades to follow. The state fixed the retirement age at sixty-five, and added significantly to social spending on the elderly. Through the 1960s and 1970s, the National Assembly had created new spaces in hospices for the elderly poor, had passed legislation on hospital reform that guaranteed the elderly a right to their choice of physicians and caregiving institutions, and had begun issuing housing allowances for the elderly as a right of retirement.[81] Yet for many little had changed. Simone de Beauvoir, for example, saw the provisions of the Laroque Commission as an insufficient security mechanism that preserved the elderly at the status of bare life. As she argued in her 1970 volume *La vieillesse* (The Coming of Age), the elderly in France were a population at the edge of humanity. Like the child, the old person was in some ways a dependent being who relied on social intervention for survival. Yet there was an important difference: "The child being a future contributor, society invests in him to ensure its own future, while in its eyes the old man is nothing but a corpse under a suspended sentence."[82] Official policy reflected this social prejudice. Echoing Laroque nearly a decade after the report's publication, she noted that "in the aftermath of the war, we made an effort to elevate the birth rate and a significant part of the budget was dedicated to family allowances: old age was sacrificed" to this cause.[83]

Beauvoir foreshadows Foucault and Agamben in eerie ways when she points to the operation of a capitalist society that relegates a population to death, letting some die as it compels others to live. Where Boverat noted that "thankfully," France could not allow the elderly to die for want of treatment, Beauvoir saw doublespeak rather than compassion. "A hypocritical decency forbids capitalist society to get rid of its 'useless mouths'," she argued. "But it gives them just enough to survive on the threshold of death. 'It's too much to die on and not enough to live on,' said one retired worker sadly. And another: 'When we are no longer capable of being workers, we are only good enough to be corpses'."[84] And while big labor always kept one eye on retirement, its major concern was future employment rather than aging. "Such as it is," Beauvoir argued, "society imposes a hideous choice: either sacrifice millions of the young, or allow millions of the old to waste away miserably. Everyone agrees that they don't want the former situation; only the latter remains. It is not only the hospitals and the institutions: it is all of society that is, for the old, 'death's antechamber'."[85]

As Beauvoir wrote, things were changing somewhat. The population was aging ever more rapidly: between 1962 and 1974, the number of retirees doubled in France, and the number over sixty-five grew by 25 percent,[86] but the amount of state aid to the elderly grew sixfold. At the national and local levels, the French introduced a range of new opportunities for the elderly. "Lifelong universities" allowed for the productive use of retirees' expanded leisure time, and "Clubs de troisième âge" provided a social space for fighting isolation. Publishers developed magazines, books, and advice manuals for the elderly, and broadcasters produced radio and television programs geared toward an expanding audience of seniors.

But while state spending improved the economic fortunes of many French seniors, and the private sector offered new social possibilities, at the levels of both experience and representation things were getting worse if anything. One study conducted in 1976 at a Centre National de la Recherche Scientifique (CNRS) institute indicated that the state's protections amounted to little more than "asylumization" in a network of hospices. Poor elderly women with no surviving family in particular were likely to end up in state institutions with medical diagnoses that marked them for social "exclusion" and "segregation" from society; worse, the authors described Paris's hospices as draconian institutions whose disciplinary atmosphere sought submission more than they delivered care, and in which suicides amounted to "a final act of liberty and the only conduct that could check the power of the institution."[87]

A literature aimed at seniors often reinforced this segregative dimension, urging them, in one scholar's words, toward "self-renunciation and abnegation." For example, the "Ten Commandments of Aging," published in the journal *Gérontologie* in 1972, offers a scathing reflection on the place of aging in the French imagination that merits citation in its entirety. Addressing the elderly in a condescending "tu" form, it treats them as childish:

1. Avoid doing things impulsively. Always act slowly. Think about what you are doing. If you let yourself be guided by your unconscious, you will do strange and silly things. Looking in the drawer for the nutcracker you need, you'll return with the corkscrew.
2. Speak as little as possible about your pain and health problems. Remind yourself that they don't interest anyone. If you speak about them, you will bore people. They will listen to you with pity while thinking about something else, and will hurry up to be rid of you. But speak to them about their own health, their children, and their business, and you will see them interested right away. They will be quite happy and will leave you thinking that you're quite a likable old man.

3. Go to see your friends, but not too often, and don't stay too long. If you do, they'll think: "He annoys me, this old man who comes every morning and wastes my time!"
4. If you live with your children, and they have friends to dinner, make the pretext that your eyes are tired in order to go to bed early. After you leave, the conversation will be more exciting, more personal, and more joyful.
5. Don't be a burden to your relatives, or at least minimize the burden. Prepare yourself so as not to need anyone. Gladly accept being alone.
6. Learn to appreciate, to love, and to take advantage of your solitude. It will allow you
   - to take long walks, excellent for your physical health;
   - to develop your cultural life, in reading books or listening to programs that interest you;
   - to cultivate your natural aptitudes: music, drawing, watercolors, etc.
7. Note from day to day, in a journal, the important moments of your personal life, the reflections that your personal encounters and your readings and thoughts suggest to you, that strike you and that you don't want to forget. As the months and years go by, this written memory will render you ever greater services and will seem more and more indispensable.
8. Go to see your doctor as little as possible. Remind yourself that he can do nothing about aging. It's you alone who can delay its effects for the longest, in struggling to maintain, despite your natural laziness, a physical and intellectual activity that will slow down the atrophy of your muscles and the sclerosis of your brain.
9. Take an interest in your neighbors and those people with whom you are in contact. Do them favors every time you can. If, because you have the time and the means, you can even do more significant things for people, do so discerningly, but generously. They don't appreciate it? So much the better! Because what counts for you, is the goodness of the intimate feeling of having helped friends.
10. Think serenely of your ever more closely approaching death, but never speak of it. Not for the sake of not making others sad. Remind yourself that they will consider your death as a natural, normal, and even (if they think you're a bit senile) desirable event. But rather because it will make them think of their own deaths, and that will be very unpleasant for them.[88]

These "commandments" appear in the journal with no context: it is impossible to tell whether they are satirical or not.[89] But even as satire, they reflect an atmosphere of relegation for the elderly. Many elderly acknowledged a widespread sentiment of the worthlessness and dehumanization these commandments suggest, as a series of longitudinal interviews with aging French citizens indicates. In the mid-1970s, a team of sociologists led by the demographer Françoise Cribier of the CNRS in Paris began an extensive study of two cohorts of retirees: one group born around 1907, and another born around 1921. The goal was to investigate the life experiences and economic conditions of those born before and after the First World War, including the first generation to retire after the Laroque Commission's recommendations went into effect, as well as those who retired after policy changes had become well established. At its peak the study enrolled over 2,000 participants, nearly all in the Paris region.[90]

The records of this study constitute an immense archive. Each dossier contains extensive interviews and surveys with the participants at roughly ten-year intervals: 1975, 1985, and 1995. These include interviews about the participants' sense of place (the team concerned itself primarily with subjects who had been born in the provinces and had moved to Paris),[91] richly detailed surveys about housing quality, and investigations of participants' plans for their end of life. But the interviews also provide keen insight into the participants' experiences of aging in contemporary France, with the most extensive data coming from the second phase of the study in the 1980s.[92] The interviews reveal a population in deep reflection over their condition on the margins of citizenship. Participants talk about aging in place in a dramatically changing world, one that menaces them with physical and social insecurity. They discuss their increasing isolation from that world, and that world's increasing contempt for them. Perhaps most disturbing, they show an internalization of that contempt: a self-hatred and a devaluation of old age as a condition at the limits of humanity.

The sense of a loss of place and tradition was paramount for many participants. For some, this meant a world that was changing around them—one usually changing for the worse. Immigration was transforming neighborhoods, for example. One woman described her neighborhood as having been "a good one when I got here but now there is, like everywhere, an influx of . . . different people."[93] Another described Paris as "not the same as before. We're no longer living in France you could say."[94] For one man, "you never hear anyone speak French anymore."[95] Others ascribed the transformation of society to a generation gap rather than immigration. One elderly couple described "the Africans and Algerians" as "well dressed" and "not a problem," instead describing French youth as "sloppy," "careless," and "not very charming."[96] The generation gap contributed, for many participants, to an atmosphere of selfishness, vandalism, and danger. For one seventy-two-year-old man, Paris "was much more friendly, more affable too" in the past; now "everything has changed."[97] One woman declared that in comparison with the past, "everyone lives for himself . . . it's colder now." Her husband agreed: "There's less camaraderie than in the past . . . these young Parisians can come, they brush right by you, they say nothing."[98] Social involution produced social isolation. One woman noted that "before we had, you know, good neighbors, I don't know, it was more familiar." But "now we withdraw more easily."[99]

The generation gap and new forms of insociability also engendered fears that in turn exacerbated isolation for many participants. "The rue Mouffetard has changed, it's all young people," noted one woman. Although she had "a new park in her neighborhood," she did not visit it because she was "afraid it will be destroyed by vandals and hoods."[100] Another said that she "wouldn't dare go to the theater at night, I'd be scared coming home."[101] And another, who had "been robbed," described herself as "afraid all the time in this house": "I don't know how long I can continue like this."[102] One woman who had been "attacked twice . . . by teens" told the interviewer, "I'm sorry that I can't go out any longer because at night, you get attacked, so I can't do anything anymore. So I have to shut myself in and watch television. That's all we have."[103]

For many elderly, disability and fear of injury increase isolation. One woman indicated that she could no longer take the metro because she couldn't negotiate the stairs. Even getting on the bus was difficult: "That's why I don't go out much."[104] Another woman said that because of the pain in her legs, she rarely left home.[105] One eighty-eight-year-old woman interviewed in the 1995 cycle said that she was "too afraid to risk" even a trip to the park because of her anxiety about falling: "I don't dare take the bus." She lived in a fourth-floor walk-up with no bath or toilet at the time, so any additional disability would force her out of her home. She desperately wanted to avoid moving into a nursing home. "Old folk's homes, I see them on TV, but that's not really true, TV, because it's not reality. They show us what we want to see. . . . I went to see a lady in one of these homes. . . . It was disgraceful."[106] Hearing problems were another disability that affected social life. One man stopped going to church because he couldn't hear the services, and said he no longer liked the company of others, because "it's tiresome to make people repeat themselves two or three times."[107] For another man, increasing deafness even ruled out television as entertainment.[108]

For others, isolation was a consequence of aging out of a dying social network. Speaking of his friends, the man who could no longer watch television said, "We see them die one after the other"; his wife commented, "It's tiring to get old."[109] One woman said that she had had friends, "but now that the husband is dead," they no longer visited. Her own husband commented, "It's funny, when we get old, we withdraw a bit."[110] Another said that "at my age, there are people who disappear."[111] Another Parisian man echoed the sentiment: "I don't have a lot of friends because you know at 76 years! There are those who are gone."[112] The death of friends and relatives disrupted long-range plans as well as social networks. One man planned to live with his sisters (one a few years older, one a bit younger) after his retirement. "I had thought I might spend the summer with them, that I'd eat every day with them, and then they both died within four months of each other." Unable to return to his career, he lived in a deep depression that added to the burden of loneliness.[113] Others cited exhaustion and fear of social rejection as reasons for their isolation: "Having people over, that's onerous and tiring"; "I have a neighbor, she's nice, she waves from the window, she says hello, but I don't dare invite her, what do you want, I'm afraid of a rejection."[114]

The participants also indicated their awareness of rampant ageism, sometimes from youths, sometimes from the middle-aged. The woman who lived in the Rue Mouffetard complained about one shopkeeper as indicative of a larger rejection of the old in contemporary Paris: "He's awful with old people. There was a woman who asked him for a paper bag, an old woman, and he said to her, 'How did you carry things in 1914?!'"[115] Others spoke of being insulted by teens on the street, and about the "demoralizing" depictions of the elderly on television.[116] This was also a product of a changing time; in one participant's view, "The young don't have the respect for the elderly like they used to."[117] More disturbing is the internalization of ageism. Many participants displayed a near self-hatred in the interviews, reflecting a broader social ageism in their self-presentation. Some echoed the "Ten Commandments of Aging," insisting a bit too strongly on how much they liked their solitude. One man said, "We get to an age when we like . . . it's not that I don't like company, I like company well enough, but I also like tranquility."[118] For others the internalization was more vivid. One man refused to "watch shows for seniors: I have enough of my own age to deal with without having to deal with that of others."[119] Another man refused to socialize with those his own age because of their constant griping about pain and suffering: "One will complain about this, another about that."[120] One woman said that she had to "accept" aging: "It doesn't do anything to cry, to moan, or to complain. Because those around us don't like the old much. They don't like the old very much at all. . . . I can't cry about it, that wouldn't help anything."[121] This could represent stoic resolve, but others' reflections indicate that such sentiments were much more an internalization of broader social representations of old age in a society that had rejected it. The participants nearly universally condemned municipal senior centers, for example. Designed specifically as state-funded social centers to combat the isolation of the elderly, the senior centers met with most of the participants' contempt. Some indicated that the senior centers could be socially exclusive,[122] but most rejected them as "demoralizing." For one woman, "to see these little old people trying to do gymnastic movements, it's enough to make you cry."[123] For another woman, eighty-eight, "to be only with old people doesn't please me at all."[124] In the eyes of one married couple, the senior centers were a symptom of elderly isolation rather than its solution: "The center, oh no, that's for those who are alone, we're two, we don't bother with that."[125] Another couple agreed, saying, "That doesn't interest us."[126]

The sentiments of the elderly who participated in this project reveal simultaneously a profound, if veiled, suffering and its disavowal. Few of the participants admitted their isolation directly, providing instead important suggestions of its depth: they mentioned the increasing frequency of their friends' deaths; the difficulty of seeing their children, now busy with their own families; steadily mounting difficulty navigating staircases and public transportation, leading to longer periods of confinement to the home; increasing fears of changing neighborhoods and environments that exacerbated their reclusion. But many indicated a deeply ingrained self-abnegation that a discourse of aging originating in the postwar era had initiated. As Cribier herself argued in 2003, since the 1970s (and indeed, as this chapter argues, since the 1940s), the French had greeted their dramatic increase in life expectancy not with celebration, but with a sense of "crisis."[127] If society embraced retirement as a part of adulthood, it did so while rejecting old age as a costly social burden. Yet while Cribier argues that a moral economy of age should trump a fiscal one, it is also clear that the language of aging cast the elderly not merely as a threat to national solvency, but also to humanity: "corpses under a suspended sentence." At the turn of the millennium, national and social rhetoric closely linked to a political economy of marginalization had pushed the elderly to the boundaries of the grievable, as well as to the boundaries of the human.

Marked for Death: The Elderly and the Heat Wave

> Depression overtakes me, but I'm used to it. I have no visitors, before I knew the old neighbors, we saw each other a bit, but the new ones I don't know.
>
> MADAME JEANNE R., 1995 (d. 14 August 2003, Clichy)

By the summer of 2003, most of Cribier's initial 1907 cohort had died: only 145 of the initial 1,370 remained. Only one of them, Madame Jeanne R., died during the heat wave. Jeanne was emblematic in many ways of those who had died. She had been born in 1912, and was ninety-one when she died on

14 August in a nursing home in Clichy. The study had lost touch with her after interviewing her in 1995. But at that stage she had begun to experience profound disability and isolation. Then in her early eighties, she lived in a tiny apartment in Saint-Denis that she had purchased with her husband in the 1930s; she had lived there alone since his death in 1972. Already in 1995 she was losing her eyesight because of cataracts that surgery had failed to correct, and was diabetic. Despite her two knee replacements, she lived on the third floor with no elevator and had no toilet or running water. She had no children, and her only social contacts were a distant niece and a social worker, without whose help she could not leave the apartment. She was also desperately poor, living below the threshold of taxation.[128] She was a typical victim in a number of ways. Her poverty, disability, age, and sex imprinted her with risk. Like many in her cohort, she was also nearly completely isolated, and had internalized this condition—as well as her poverty—as her natural state. With no close relatives or friends, she would almost certainly have found her end at Thiais among the forgotten had she lived in Paris rather than Clichy.

The one condition that rendered her somewhat less typical was her residence in a nursing home at the time of her death. Only about one in five of those who died during the heat wave died in nursing homes.[129] Among the forgotten victims buried at Thiais, the number is even lower, with only three of ninety-five living in such institutions at the time of their deaths. There are a number of explanations for the relatively lower institutional death rate. One is the closer medical monitoring of institutionalized patients: in contrast to the living situations of those who live alone, nursing homes and assisted living facilities have established structures for attending to their residents, so they are less likely to die preventable deaths such as those due to heat stroke or dehydration. The epidemiological literature supports this conclusion, noting, for example, that in nursing homes during the heat wave, bedridden patients had better outcomes than did ambulatory patients precisely because they received closer monitoring and were also more likely to be on intravenous hydration therapy.[130] And as those with few social contacts, the forgotten—perhaps ironically—were less likely to find themselves placed in a long-term facility in the first place. One important counterexample is Raymonde, buried at Thiais, who died in a nursing home in the nineteenth arrondissement on 10 August. Raymonde had lived in the nursing home since 1991, and was in generally good health and went out on most days, but was "frightened of dying" and was "fragile," according to one nurse with whom I spoke. The nurse added, "She could not tolerate the heat." Raymonde had two daughters who were in frequent contact with her, but they were away on vacation when their eighty-nine-year-old mother died. Through a bureaucratic error, Raymonde ended up at the city morgue on 12 August, where overwhelmed staff placed her with the forgotten, leading to her burial at Thiais. The daughters were furious, and began a long legal fight to secure their mother's removal and reburial in a private cemetery plot. The case is important because of its unlikeliness: Raymonde found herself in a good facility precisely because of her family links; it was despite those links that she ended up among the forgotten at Thiais.

Yet even though nursing homes reduced risk during the heat wave (somewhat—they still witnessed a staggering 2,500 deaths), their conditions remain an important symptom of the relegation of the elderly in French society. The Laroque Commission proposed an expansion of the nursing home system, which it recognized as inadequate and ineffective for an expanding aging population, and also proposed a reform of the system, which the state enacted in 1965.[131] But even with this state endorsement, conditions remained grim inside the hospices and other institutions. In 1970 Simone de Beauvoir described nursing homes as sites of "a literal medical abandonment. . . . The bedridden old end their lives there in an apparent general indifference."[132] Since Beauvoir's indictment, social knowledge has made the nursing home a metonym for old age itself—as an entryway into death's waiting room. As one journalist put it in 2003, "To be confined here is already to die a little."[133] Their residents are "citizens at the end of their rights," one sociologist wrote in a 2002 exposé: even their personnel became dehumanized through a "daily habituation to decrepitude and death."[134] For the poorest of the elderly, these homes are, according to one critic, "neither camp nor prison. In a sense, they are worse." Here "one dies of the deep sadness of death, without tragedy. . . . Here, nothing but the end, nothing but the end."[135] The language here evokes the degree of abandonment that the anthropologist João Biehl documents in his chronicle of the life beyond citizenship of the mentally ill and HIV-positive homeless in contemporary Brazil, whereby state and community discard the socially unwanted into an antechamber of death.[136] Placement thus initiates an abandonment of the "good life" in an Aristotelian sense—the political life, the life of citizenship and the individual—for the bare life of invisibility, an act that amounts to a de facto abandonment of personhood and the right to the good death, the death as individual that contains meaning for family and community: in short, Paulette's death in *Ensemble, c'est tout* as opposed to death on the trash heap of Houellebecq's dystopian fantasy.

The rhetoric of aging in the postwar era has linked devaluation with dehumanization. Claims about the economic threats of aging have accompanied a negative biopolitics—a thanatopolitics—that has at least implicitly suggested death as a technique for influencing population: Boverat's "Could we let them die?" of 1951. The heat wave cast this problem in high relief. The expenses of managing the health of an aging population have steadily increased in France.[137] By the end of the 1980s, one journalist went so far as to declare the elderly "gluttons" of the health-care sector, more interested in preserving life at the limits of existence than in the solvency of the state.[138] In an example of particularly bad timing, the French finance minister Francis Mer announced at the end of July 2003 his "solution" to this "problem of health insurance" costs in France: "Quite simply, get rid of the last year of life, because that's what costs the most for social welfare."[139] Within two weeks, this vision had come true. Given the high death toll among the elderly from the heat, the satirical newspaper *Le Canard Enchaîné* declared that the Ministry of Health should be renamed the "Ministry of Economy."[140]

Health Minister Jean-François Mattei erred arguably even more greatly. As noted in chapter 1, on 11 August, aware of the crisis, he appeared on the news from his vacation home, urging calm and dismissing the numbers at a few hundred excess deaths of those who likely would have died soon anyway. On 14 August, upon his return to Paris, he indicated how little value the elderly held as a constituency. When a reporter asked him why early warning systems for heat risk were not in place, he replied, "You know, the elderly, they don't have very good memories, often from one moment to another, so the preventive messages that we could air . . . well, they'd forget them the same day!"

These sorts of sentiments write the elderly poor out of existence as fully human individuals. Isolation of the elderly is an important social reality, one often compounded by poverty. According to recent census data, this is particularly a problem for aging women. Whereas among Parisian men sixty-five and older, 70 percent are married, only a third of Parisian women sixty-five and older are married. As the population ages, the difference is even starker: nearly two-thirds of men eighty and older are married, compared to less than a fifth of women.[141] And the economic circumstances of the elderly, although improved since Beauvoir's day, remain scandalous. According to recent data, some 10 percent of those living below the poverty line are above sixty-five.[142] Above age seventy-five, women are half again as likely to live in poverty (12.5 percent) as men (8.5 percent).[143] Living in isolation explains much of this disparity: elderly women, far likelier to live alone than elderly men, cannot benefit from the economies of scale that life as a couple or in a family provides, thus stretching their already smaller disposable income further than do men in the same age group.[144]

Eugénie offers a useful case for thinking through the realities of isolation and poverty. At eighty-seven, she had lived for as long as anyone could remember in her apartment in the Boulevard Poniatowski in the twelfth arrondissement. Eugénie had a daughter, the *gardienne* and neighbors told me, but she had never visited, in their recollections. "She was really all alone," the *gardienne* insisted. She had once had some financial means, but gradually her resources began to dwindle precisely when she began to suffer from increasing disability. Her failing sight had left her "practically blind," her next-door neighbor told me. Plus she suffered from leg problems. Her neighbors, quite elderly themselves when I interviewed them in 2007, said that they had tried to do what they could for her. They did shopping for her and ran small errands when they could. But she had stopped leaving her apartment at all, relying exclusively on the charity of her neighbors and of the grocer on the ground floor of the building, who delivered much of what she needed. One of her neighbors told me, "We were nice to her, because she deserved to be liked." But "she lived on very little." She had no social assistance, no *aide ménagère* to help her with day-to-day tasks. But with no one to discover her—not even a social welfare aide—she remained in place for more than a week after she died. It was only the stench in the hallway that alerted neighbors to her death. When I visited in 2007, there was still a mark on the door from where health authorities had sealed the apartment until a disinfection team could sanitize it (fig. 21).

Cases like this are plentiful. It was only René's domestic helper, appointed by the city, who discovered his body in the sixteenth arrondissement after

FIGURE 21. Eugénie's (d. 20 August 2003) door, twelfth arrondissement. In 2007, the door still bore a stain from the police seal closing the apartment as a scene of investigation. Photo by the author.

he had been dead for days. And just around the corner from Eugénie, in the Rue Marcel Dubois, Petar, a seventy-seven-year-old Serbian who had lived in Paris for some twenty years, was one of two isolated seniors who died in the same building. According to the *gardienne*, Petar only ever left his apartment in the evenings, and even then, not often. He was quiet, an ideal tenant. But in mid-August, the tenant who lived directly beneath him complained of a leak in his ceiling as he left for vacation. The *gardienne* knocked on the door repeatedly with no answer. As she had no keys to the apartment, she did not enter, but more than a week later, the smell of Petar's death pervaded the building. Authorities who signed his death notice on 27 August indicated that his body had been decaying for more than two weeks. Another tenant in the same building, an elderly woman, also died during the heat wave. This one the *gardienne* discovered herself when she entered the apartment with first responders. She was in bed, wearing a sweater, under a blanket and a comforter. "She was cold!" the *gardienne* told me. But she also said: "I will never forget the skin of someone who has died from dehydration. It was like bark on a tree."

These cases attest to a broad social problem of isolation—an epidemic of elderly vulnerability in a period of extreme crisis. But following the dehumanizing rhetoric of Mattei and others, isolation is less a social problem than it is a function of a growing disablement that excludes the elderly from full membership in the community. According to this logic, such isolation is therefore inevitable, rendering their deaths also inevitable and merely hastened by extreme temperatures: the biology of aging in modern culture has opted the elderly out of the social contract. Their aging, poverty, and isolation constitute transgressions against citizenship. They exist in a state of exception—as Beauvoir said, in death's waiting room—outside the domain and protections of the human polity. Their lives are those of social abandonment, effectively "invisible," as one journalist put it after the heat wave.[145]

Historians and social scientists have traditionally cast this problem in economic terms. Indeed, the labor and economic histories of aging have been central to the story of contemporary France. Since the *trentes glorieuses* the nation has undergone turbulent economic and social change. The oil shock and inflationary crises of the 1970s, the immigration reforms of the 1980s, and welfare reforms of the 1990s and 2000s all made significant marks on the ways in which the French conceive of the relationship between economic concerns and social citizenship. The pension reforms of the 1990s in particular precipitated major social upheaval. The 1993 reforms of the Édouard Balladur government raised the labor-participation rate from 37.5 years to 40 years for a fully vested pension, while also raising the pension rate from a ten-year high average to a twenty-five-year high average for the private sector; the 1995 reforms of the Alain Juppé government attempted the same for public employees. The result was a weeks-long general strike that paralyzed the country, in the course of which many on the political left decried the dismantling of the welfare state. In a 1995 speech to a crowd gathered at the Gare de Lyon, Pierre Bourdieu called the move a step toward the "destruction of a civilization" that was itself "connected to the existence of a public service and to equal rights in a republic," while a statement signed by a number of prominent French intellectuals defended the strikers against a "government offensive" that threatened "the most universal achievements of the Republic."[146]

On the other side, many have found such reforms essential to rein in runaway government spending on the elderly, whom they have signaled as an overprotected class. Echoing the journalist who called the elderly "gluttons" of the health-care sector, the historian Timothy B. Smith has noted that by 1985 the elderly consumed some 70 percent of social spending in France. He argues that this is the result of a model that attempted to correct the desperate poverty of the generations of the two world wars and the Depression. Yet this system has remained on the books, despite having "wiped out old-age poverty" in the 1960s and 1970s: as a result, he argues, retirees now purchase some half of all first-class travel tickets and new cars in France, largely on the public dime.[147] Meanwhile, public spending on those whom Smith claims truly need it—namely, the young, through family allowances—has declined relative to population. The Balladur reforms were essential, he argues, but even they leave the so-called welfare generation—those who benefited from both economic growth in the postwar era and a generous welfare state—relatively protected.

This chapter indicates that both sides missed the mark in this debate. Hand-wringing on the left about the destruction of the welfare state and civilization itself through a reduction in pensions represents an illusory nostalgia for a golden age of retirement that never truly existed for many of France's poor—in many ways, a correlate for the nostalgia for a traditional France that is at the core of the narratives in the Cribier studies.[148] Despite dramatic increases in pension spending since the 1940s, many of those who most needed the state's help found themselves in desperate circumstances not only in 2003, but in the preceding decades. But for critics of the welfare state, the heat wave's revelations of elderly destitution belie the idea that increased pensions "wiped out old-age poverty." Data may suggest that contemporary French retirees are in the aggregate better off than the average French person, and are indeed the country's best-protected generation. Yet a tour through the dilapidated halls of the heat wave's forgotten dead indicates that such a frame is rhetoric rather than reality for many for whom the state's protections have constituted a state-sponsored preservation of bare life rather than a retirement package. Muguette, Georgette, Jeanne, and Eugénie did not book first-class travel tickets—indeed, they could not afford to book their own funerals.

This is one reason that the notion of bare life and a broader literature on biopolitics offer a useful frame for imagining old age in contemporary France. For Foucault, aging and its social knowledge constituted a dividing practice that measured inclusion and exclusion by reifying social phenomena as biological expressions; discourses of aging thus operated in a manner analogous to those surrounding insanity, medical pathology, and sexuality in the same period. Yet whereas Foucault's investigations of other dividing practices have spawned enormous scholarship, historians have been relatively slow to engage with aging, leaving critical questions about the culture of aging in France underexamined.

Yet social knowledge about aging is clearly historically contingent. Between the 1960s and 1980s a new discourse of old age emerged in France, of which the Laroque report, Beauvoir's *Vieillesse*, Foucault's outline of biopolitics, and Cribier's longitudinal surveys are both artifacts and sources of evidence. Capital, state, and society took a significant interest in the problem of aging for the population, linking economic implications to a devaluation of life. Recent interventions regarding the concept of bare life—of existence at the margins of life, at the margins of citizenship, at the margins of representability in all its senses—suggest ways in which we can consider aging outside of its strictly economic and physiological domains, and in a more broadly cultural one. As this chapter has argued, radical transformations of postwar France have had little effect on perceptions of old age, which remain tied to an exclusionary rhetoric that dates to the eugenic moment of the late nineteenth and early twentieth centuries. Where ethnic, racial, religious, and gender exclusions have been the sites of raging public debate since the 1980s, old age has remained the site of a consensual alterity.[149] The humanist crisis that the heat wave engendered prompted a desire to imagine a possibility of old age like that of Gavalda's Paulette. But the grotesque realities of Mattei's and Mer's language, and of a broader language of abnegation of the elderly in contemporary France, share much more with Houellebecq's framing of old age as "the object of unanimous disgust."

5

Counting the Dead: Risk and the Limits of Epidemiology

Among the first cases I investigated when I began this project was that of Suzanne. The headstone at Thiais indicated that she had been eighty-three years old at the time of her death. At this point I had not yet received death notices from the city's twenty vital statistics offices—one for each *arrondissement,* or residential district—but was eager to begin investigating the victims' lives. I began by going to the Bibliothèque Nationale with a list of the names of those buried in divisions 57 and 58 at Thiais, and consulted a telephone directory from 2002 in the hope that it might contain at least some of their addresses. Although this was a rudimentary method at best—many of those buried at Thiais did not have telephones, so they remained unlisted; others had common names, so there were duplicates—it yielded significant results: I obtained the addresses of well over half of the subjects in this way, and later confirmed them with the death notices.

There were four listings that shared Suzanne's name in the 2002 directory. One was in the Square Rosny Aîné in the thirteenth arrondissement, near the Porte d'Italie on the southeastern fringes of the city. The group of ten-story apartment buildings that comprised the project dated to the 1950s. They were stucco, aluminum, and glass buildings whose end units had floor-to-ceiling windows with no shutters that spanned the width of the apartments. Conversations with the building's *gardienne* and a few neighbors confirmed that Suzanne had lived in one of these end units on the fifth floor; the apartment would have baked in the summer heat in 2003, given its southern exposure. Suzanne had owned a dry cleaning shop located on the ground floor of the building from the moment of construction to the mid-1970s, when she retired. She had never married and lived largely on her own, although she occasionally housed foreign students for short periods. According to the *gar-*

dienne, a student (Iranian, she thought) had been living with Suzanne at the time of her death, which took place at the nearby Bicêtre Hospital.

After interviewing the *gardienne*, I encountered an elderly woman walking a small dog outside the building. When I asked if I could speak with her about Suzanne's death during the heat wave and the circumstances of her life, she was extremely accommodating. We spoke for nearly an hour, and she confirmed much of what I had imagined. She had lived next door to Suzanne since 1956, and she knew her quite well. The neighbor was ninety-one when I interviewed her in 2007; she had been eighty-seven, a few years older than Suzanne, during the heat wave. She told me that Suzanne was not in good health, and that she left the building increasingly infrequently. Her only social contacts were the students who lived with her and a friend who came to visit fairly often.[1] The neighbor then told me about her own experience. She said that she did not feel in danger during the heat wave, but instead merely felt "too alone." She had a daughter, but had basically lost contact with her. "It's not fun being old and alone . . . it's tough," she told me. "They don't do anything for people who are alone; the city should do more." Gesturing to her Yorkshire terrier, she said: "My little dog is already fifteen years old. If he dies, I won't live much longer."

The neighbor's account of the heat wave and of aging in place was fascinating. It suggested the possibility of conducting what epidemiologists call a "case-control" study: with one subject dying during the heat wave but a next-door neighbor who shared many characteristics with the victim—age, socioeconomic status, sex, marital status—surviving, this coupling offered a useful comparison. The neighbor talked about the pressures of aging in a neighborhood that witnessed increasing levels of crime. She discussed the case of another neighbor—also an elderly woman—who had been attacked in the building, and told me that even on the fifth floor, she would occasionally hear intruders trying to open her locked door at night. The critical factor that led to her survival appeared to be her pet: walking the dog three times a day got her out of the building and forced her into some social interactions with those around her. Indeed, during our discussion several neighbors stopped to speak with her.

But the encounter also revealed the pitfalls of field interviews. My discussions with several people about Suzanne's life and death yielded a great deal of information. Yet several weeks later, I was stunned when I received Suzanne's death notice from the Bureau of Vital Statistics in the thirteenth arrondissement. The date of death listed for the Suzanne at this address was not 4 August 2003, as the headstone read, but was instead 14 October 2002, ten months *before* the heat wave. The date of birth was not in 1920, but instead in December 1908, making her not eighty-three but ninety-four. By trusting a telephone directory I had located the wrong Suzanne.[2] But at least three who knew the woman who had resided in their building were perfectly willing to attribute her death to the 2003 heat wave. Several factors could explain this misunderstanding. It had been four years since the heat wave, and Suzanne had died within a year of the disaster. The details may have become somewhat less reliable during the interval. Also, my line of questioning certainly biased the informants' responses, as my questions linked Suzanne's death to the heat wave perhaps too insistently. Yet even the neighbor who had lived alongside Suzanne for nearly fifty years appeared to believe in the course of our conversation that Suzanne had died from the heat.

This chapter argues that such misattributions are easy to make, given the power of the epidemiological profiling that characterized investigations into the heat wave. In the disaster's aftermath study after study confirmed that the elderly constituted the bulk of the victims, and that the elderly were therefore the primary population at risk for death by extreme heat. Media responses to the disaster included a wave of stories about aging and isolation in France. The typical heat wave victim, according to extensive demographic studies and popular representation, was a poor, isolated, elderly woman not at all unlike Suzanne. It was easy to associate heat wave deaths with the elderly given the pervasiveness of this image as well as official prevention policies that followed the disaster, which aimed almost exclusively at the elderly as a population at risk (see figs. 22 and 23). Yet some 3,000 nonelderly victims also died during the heat wave, and an exclusive focus on the elderly as the principal risk group has rendered their deaths invisible. As I describe below, my fieldwork illustrated this phenomenon with devastating clarity. While it was easy to assume a figure such as Suzanne had died of the heat, my informants frequently refused to consider the heat a principal cause of death for younger victims, attributing their deaths to alcoholism, obesity, a heart condition, addiction, mental illness, or another affliction.

A willingness to consider the elderly at risk for heat-related deaths while simultaneously overlooking the heat as a causal factor for a younger age group betrays a critical weakness of epidemiological profiling—one signaled as early as the mid-nineteenth century by one of the field's pioneers. An aggregate picture of mortality risk during the heat wave highlights the elderly as the primary vulnerable group. Yet an exclusive focus on the elderly overlooks the circumstances of the others who died during the heat wave. Aggregate profiles model risk with great accuracy and have the capacity to direct policy and funding in the most useful direction. But they are also blunt instruments that run the risk of blinding the state and the community to other critically

FIGURE 22. Poster for campaign to promote awareness about heat risk in Paris, summer 2007. Photo by the author.

FIGURE 23. A similar poster from the same campaign. Note the child aiming the fan at the elderly man in the photo. Photo by the author.

vulnerable groups. This chapter examines the tensions between the aggregate models of risk that epidemiologists constructed in the aftermath of the heat wave and the particular cases of certain victims whose circumstances expose the limits of standard practices. By exploring the methods by which demographers and epidemiologists tabulate the impact of disasters, and by investigating the deaths and life circumstances of a handful of the heat wave's forgotten victims, it discusses the ways in which the counting of the dead—and the attribution of causes of death—often aggregate risk through normative, exclusionary, and ultimately divisive practices, with grave potential consequences for understandings of risk and resilience in periods of disaster.

Who Is a Disaster Victim? Models of Risk and Death without Cause

According to official reports, the Haitian earthquake of January 2010 killed 316,000 people. The death toll from the Tohoku tsunami in Japan in March 2011 currently stands at 15,844, but radiation deaths and deaths from cancers linked to the associated nuclear disaster at Fukushima Daiichi will surely elevate those figures over time. Hurricane Katrina killed 1,836 in the American South in 2005, while the Sichuan earthquake in 2008 officially took 69,197 lives. And in late September 2003, demographers and epidemiologists working for Inserm stated that the August 2003 heat wave killed 14,802 in France.

States and societies rely on two principal measures for gauging the suffering that disasters cause: death tolls and insured losses. These are critical figures, as they often tell us a great deal about both the nature of the catastrophe and the environment in which it took place. But such measures can be misleading. The Loma Prieta earthquake in California in 1989 and the Haitian earthquake in 2010 both measured 7.0 on the Richter scale. Measured by mortality, the Loma Prieta quake caused 63 deaths versus the 316,000 in Haiti; in economic terms, the Loma Prieta quake resulted in over $6 billion in property damage and $960 million in insured losses. There were almost no insured losses in Haiti—a consequence of a near-total absence of property insurance on the island.[3] The geological marker suggests at least a rough parity between the quakes, while the mortality associated with each indicates that the Haitian earthquake was far more severe. The costs associated with property damage and insured losses are more a gauge of property values and the level of economic development in the San Francisco Bay area than they are an indication of the severity of the disaster.

These figures help us to assess the magnitude of disasters. It is difficult to think of disasters—much less to define or to rank them—without also imagining them in terms of their death tolls or other damages they cause.

But where do these numbers come from? To what extent are they estimates? How rough are they? Do they include mortality indirectly associated with a given disaster, or only direct casualties? And on what geographic and temporal scales? Different methodologies yield dramatically different tolls.[4] The numbers associated with disasters serve important political, emotional, and economic purposes. These are therefore important questions, and the answers to them reveal enormous uncertainty as to who counts as a disaster victim, and ultimately, what constitutes a disaster.

A case such as the disaster at Chernobyl casts these uncertainties in high relief. When a nuclear reactor exploded at Chernobyl in April 1986, workers who struggled to contain the damage suffered inordinately high radiation exposure. Roughly a hundred died in the weeks that followed. In the long term, thousands more have suffered deaths associated with radiation exposure and cancer. Levels of thyroid cancer among children have been particularly devastating. But the disaster indicates how difficult and how contingent mortality estimates can be. If measured in the immediate aftermath of the explosion, the death toll is about a hundred; if measured in decades, it could reach well into the thousands.[5] The murkiness of these estimates follows from the uncertainties of toxic exposure on human health. But even acute natural disasters yield uncertain tolls. The 1,836 who died during Hurricane Katrina of course include those who drowned in the flooding. But it is unclear whether that figure also includes those murdered in the violence that struck New Orleans after the weather cleared but as the city struggled to return to normalcy. It certainly excludes long-term mortality. Most would likely agree that someone who lost family and property during the hurricane, who was displaced in the evacuation, who failed to build a new life in Houston or Galveston, and who turned to suicide a year or a decade later was a victim of Katrina; but this hypothetical figure will never count in any official toll.

The measurement of mortality in epidemic or emergency periods has long been a controversial subject. Efforts to understand fluctuations in death rates find their origins in the beginnings of biostatistical research. As early as the 1720s, the English physician James Jurin pored over decades of death records attempting to identify the likelihood of a Briton dying from smallpox (1 in 7), which he then compared to the likelihood of death by smallpox variolation (1–2 in 100): such work became a foundational defense of the preventive practice.[6] Similar debates raged in France—although a bit later—with Daniel Bernoulli's statistical arguments in favor of variolation offering a powerful contrast to Jean d'Alembert's impassioned argument for individual liberty with regard to the risks of the practice.[7] Michel Foucault argues that the statistical measurement of death itself first emerged in France in the late eighteenth century; indeed, he considers this the foundational science of a nascent biopolitics of population. Whereas the counting of death in epidemic periods had preoccupied sovereign powers since the great plagues of the medieval period, Foucault tells us, states now became preoccupied with *endemics*, or the more or less permanent causes of high mortality in different sectors of the population. An effort to manipulate the mortality rate became the foundation of a new hygienist practice of public health: the state "had no control over death, but [could] control mortality" through a range of interventions designed to boost the birth rate and to reduce the death rate.[8]

A number of historians have corroborated Foucault's timing, arguing that a recognizably modern demography emerged in the waning days of the ancien régime. The protodemographer Jean-Baptiste Moheau published a major population study in 1778, followed by dozens of others through the end of the regime and into the Napoleonic period. Such works attempted to navigate a desire to find universal laws that governed life patterns in the population, while increasingly recognizing the importance of particular trends that varied by place, age, sex, and socioeconomics.[9] But the determination of population was highly inexact, with mortality among the most inaccurately measured components. As William Coleman has noted, the French regime estimated rather than tabulated the population, extrapolating mostly from parish records. Yet parishes typically recorded baptisms rather than births, meaning that many infants who were born but did not survive to baptism passed unnoticed by demographers.[10] In the first third of the nineteenth century, the secularization of the state took record keeping out of the parishes and placed it with the newly established *bureaux d'état-civil*, or public records offices, which came to function as the official gatekeepers of municipal vital statistics in the new nation.

The new availability of such data fostered the emergence of an applied demography that came to characterize the nineteenth-century French state. Aided largely by these records, the period—with its constant struggles against epidemics such as cholera and endemic diseases such as smallpox, typhus, and especially tuberculosis—witnessed the beginnings of a massive, highly interventionist hygienism that sought first to know mortality and then to control it through the eradication of disease, poverty, and misery. Although such work was often ineffective—with the notable exception of enormous projects such as the rebuilding of Paris under Haussmann's direction—the studies that prompted the new hygienism became a foundation of modern epidemiology and demography.[11]

The critical architect of the new study of mortality was Louis-René Villermé. As noted in chapter 3, in the 1820s Villermé engaged in a study of all-cause mortality that generated an economic geography of life and death in Paris. Villermé drew on data collected by the staff of the statistical director of the Paris prefecture, Frédéric Villot, who later became the founding director of the Bureaux de Statistique et Archives in 1822, an earlier version of France's *bureaux d'état-civil.* The project was to determine whether an apparent disparity in mortality between rich and poor neighborhoods in the city was due to accidental causes, normal statistical fluctuations, or what Villermé called "permanent causes." The study indicated that all-cause mortality was in general far higher in neighborhoods deemed poor than in those deemed rich over a five-year period, militating in favor of a "permanent cause" explanation. Villermé's team then dissected the data, searching for potential causes that explained the disparity. Geographic factors such as proximity to the river, wind direction, elevation, water quality, and population density played no role, leaving only economic inequality as the critical variable, pointing toward a political economy of mortality in early nineteenth-century Paris.[12]

As Joshua Cole has noted, Villermé's work consisted of several important degrees of aggregation. First, as had Villot, Villermé assumed a high degree of homogeneity in the neighborhoods he compared. Second, Villermé ignored specific causes of death, instead focusing on broad social trends and correlations. For Villermé, economic inequality was a "principal" cause of mortality disparities, by which he meant an ultimate, rather than proximate, cause. Although wealth and poverty were not immediate protectors of life or causes of death, they did influence general trends of resilience or risk in a population. His work therefore had predictive value only at the aggregate, collective level and not for individual citizens. Mortality was now a factor—in conjunction with economics, geography, and social position—that produced social class.[13]

But social aggregation through statistical modeling was a controversial practice. When the Belgian mathematician Adolphe Quetelet proposed the concept of "the average man" in the mid-1830s—a notion that subsumed the individual into a morass of broad population trends and probabilities that had predictive power—physicians in particular contested the usefulness of such a practice for understanding or promoting health. Paris medicine of the early nineteenth century had been founded on the careful notation of lesions and symptoms, with an emphasis on the importance of individual case histories.[14] The absorption of what physicians called "minority facts" into a law of averages threatened both clinical expertise and an emerging liberal concept of the individual: the homogenization of uniqueness presented frightening parallels to the mass democracy that threatened the liberal state in a period in which the revolution of 1830 remained seared into bourgeois consciousness.[15]

For different reasons, even Villermé called some of his own tendencies into question later in his career. As Cole argues (echoing Coleman), Villermé's 1840 study of textile workers found him reasserting the importance of the individual, rather than the group. Where means and aggregates characterized his surveys of mortality in Paris neighborhoods as well as his epidemiological studies of cholera mortality in the 1830s, he now emphasized the importance of face-to-face encounters with workers in furthering the understanding of how economic inequality shaped social experience. He writes in the study of his direct observation of workplaces, households, family lives, and meals, and how such observation could illuminate statistical knowledge.[16] This work amounted to forsaking statistical aggregation for shoe-leather epidemiology: one of the major proponents of quantitative methodology now sang the praises of qualitative approaches, with all their messiness and subjectivity.

The legacy of Villermé's earlier work far outweighs that of his later approaches. Statistical methods reign supreme in contemporary epidemiology and demography, with far less emphasis on the importance of qualitative work in the field. Fieldwork itself remains a minor industry in contemporary epidemiology, with most studies drawing on precisely the same types of data that Villermé used in the 1820s. Of the official studies of the 2003 heat wave, the lion's share employ data from the 36,000 *bureaux d'état-civil* or the national aggregator, the CépiDc (the Centre d'Épidémiologie sur les Causes Médicales de Décès) located on the outskirts of Paris. There are good reasons for this emphasis on medical arithmetic and statistical modeling. They make possible a large-scale analysis that can simultaneously examine national, regional, and local trends, with multiple codings that allow the illumination of demographic and epidemiological trends by sex, age, location, and socioeconomic status—or combinations of any and all of these—at a glance.

Yet illumination also possesses an obscuring force. As Robert Proctor argues in his introduction to the volume *Agnotology*, the expansion of knowledge in one domain can entail a concomitant expansion of ignorance in another. Even as aggregation and quantitative analysis cast certain patterns in high relief, they render others invisible. An aggregate portrait of risk that emerged from the heat wave suggested—rightly—that the very elderly, and very elderly women in particular, were the likeliest to die during the catastrophe. But even though no epidemiologist would deny that other groups, including the homeless or people with certain disabilities or illnesses, might be equally vulnerable, a consequence of aggregation is that protection of the elderly now frames the national reading of heat risk. This is of course a noble goal. But it is an incomplete one, as it leaves the entire population of the nonelderly—some 3,000 victims of the heat—out of policy debates on the social protection of vulnerable citizens. As Villermé realized late in his career, qualitative research and attention to particular cases provide a necessary complement to the essential work of quantification.

The attempt to gauge the human toll of the heat wave reveals how chaotic the practice of assembling data can be in a moment of crisis, as well as the problematic nature of a reliance on quantification as a measure of disaster. During the heat wave, counting the dead proved a controversial task. Lucien Abenhaïm, Director General of Health during the period, describes in his memoir of the heat wave a "fear of the numbers."[17] As the disaster unfolded, discussion of the death toll indicated that no one really knew what the extent of the damage was. On 10 August, the emergency physician Patrick Pelloux spoke of about fifty deaths in the Paris region in several days. The following day, when asked for an estimate of the death toll during a television interview, Health Minister Jean-François Mattei replied: "Honestly, I can't respond. I hear guesses: 50 here, 300 there."[18] On 12 August, the director of the Paris hospital network, Rose-Marie van Lerberghe, noted a hundred deaths in the Île-de-France, while the next day, Abenhaïm himself reported an estimate of 3,000 dead. On 14 August, Mattei cautioned against any estimates, noting that "we don't have any definitive numbers, we need to remain cautious." But he then immediately offered tentative estimates of 1,500 to 3,000 deaths.

The media exhibited an obsession with the body count, pressing officials at all levels to provide some means of quantifying the disaster. TF1 reported that such figures were difficult to establish, given the already fragile state of most heat wave victims. The station indicated that national mortality figures remained unavailable, but reported on anecdotal claims to surmise a dramatic elevation of mortality: Paris's ambulance services reported a tripling of mortality over the previous year, for example, while emergency services in Seine-Saint-Denis reported a doubling of the average mortality rate. But estimates reported in the media fluctuated wildly, from 100 to 5,000, leading many to question their veracity and their source.

Several factors influenced the enormous uncertainty that surrounded the death toll. Most important, no central authority had any definite figures. French law requires next of kin to register deaths at municipal offices within twenty-four hours. But the movement of death certificates from the municipal to the national level takes months at best.[19] Even had the Health Ministry collected local reports of total mortality, without cause of death information there would be no way to distinguish a heat wave victim from any other death. Death certificates routinely omit cause-of-death information, and even when they contain such information, a range of confounding factors clouds rather than elucidates the problem. Only when someone dies in a hospital with an abnormally high core temperature in the absence of an infection can a physician say for sure that the death is a result of heat stroke; this is therefore impossible to determine in deaths that occur outside of the hospital.

Uncertainty about the death toll resulted mostly from the fact that those best positioned to count the dead were far too busy either trying to save lives or processing the dead to produce an accurate figure. For example, when Rose-Marie van Lerberghe, the Paris hospital association director, offered an early estimate of a hundred deaths in the Paris region—a figure that underestimated the region's death toll by a factor of ten at that point—she had been speaking off the top of her head.[20] She justified her action later by arguing that a concern with the dead came at the expense of the living: "Frankly, I had advised [my staff] not to block the functioning of the establishments, which had other things to do than count bodies. . . . At the press conferences, we tried . . . to convince journalists how they could help us disseminate messages of prevention in order to avoid the overcrowding of the services of AP-HP [Assistance Publique-Hôpitaux de Paris, or the Paris Hospital System], but they were more focused on the number of dead than on preventive messages."[21]

As the temperature dropped, and the number of new bodies declined early in the third week of August, it was clear that thousands had died. In press conferences and interviews for the next few days, Mattei continued to suggest a range of 1,500 to 3,000 deaths. Abenhaïm insisted that 3,000 might be a minimum; on 16 August, TF1 confirmed this as the "official toll."[22] But on the same day, InVS offered a figure of 6,000–7,500. Mattei responded to these figures during a television interview by pointing fingers at his subordinates: "My staff have given an estimate of 3,000 on the high end for the period from 6 to 12 [August]. . . . Yesterday, someone gave a figure of 5,000. That's one hypothesis, it's plausible, but it's only a hypothesis, we need to clarify all this."[23] The tone of the interview was that he was only citing numbers that his staff at DGS had provided, and any inaccuracies were the directorate's problem, not the ministry's. The immediate consequence of this recrimination within the Health Ministry was the resignation of Lucien Abenhaïm on 18 August, widely portrayed in the media as an act of scapegoating.[24]

On 20 August the nation's funeral directors' association made a stunning announcement. Based on their estimates in comparison to mortality figures in previous Augusts, they had calculated the overall death toll for the heat wave at 13,000. They arrived at the figure based on extrapolations from the volume of deaths handled by public funeral services, which normally account for 25 percent of the market. The news was shocking for two reasons. First, at double the highest figure that had circulated from state officials to date, the funeral directors' estimate was staggering. Second, the number came from an unlikely source. Funeral directors have no epidemiological or sentinel function in French public health circles, and yet it was this group that provided the first solid figures on the catastrophe.[25] In subsequent days the state gradually confirmed these numbers. By the end of August the Health Ministry announced a preliminary estimate of 11,435 excess deaths for the first two weeks of August. (This mirrored the funeral directors' estimate, which included the entire month of August.) Mattei also ordered an official investigation into the death toll by France's leading demographers, commissioning Denis Hémon and Eric Jougla of Inserm, France's national health research institute, to lead the study and to produce definitive figures. When they released their report on 25 September, they had calculated an official toll of 14,802 deaths. This figure established the heat wave as the worst natural disaster in French history, having more than doubled normal mortality rates in several particularly vulnerable regions.

The uncertainty regarding the death toll resulted from a broader uncertainty about how to attribute death to the heat wave. There are several ways to measure death during a heat disaster, each with significant flaws.[26] The most conservative measure is to count the dead based on death certificate data. French death certificates contain both demographic information, including age, residence, marital status, and birthplace, and medical data, including cause or causes of death. By this measure, only those with a cause of death directly linking the death to heat (heat stroke, dehydration, certain forms of heart or kidney failure) would count as heat deaths. Such a measure produces an exceedingly low estimate of mortality by definition. Heat stroke, for example, is a major cause of death during heat waves. But the medical definition of death by heat stroke—a core temperature of 41°C (105.8°F) at time of death—means that only those who die under medical supervision (with their body temperatures recorded as they die) qualify for that official cause of death. Virtually none of those who died in their homes can figure in this model. In addition, even though French law mandates that physicians list a cause of death when signing a death certificate, roughly a third of all death certificates in France include no cause of death. In periods when excess death places enormous stress on the mortality system, such lapses are likely to increase.

Another method for counting the dead—and the one that French demographers used to assess the mortality linked to the heat wave in 2003—errs in the other direction. The measurement of excess mortality is a simpler and blunter tool than a tabulation of death certificate data, but one that probably provides a closer estimate of the real death toll. It involves counting the number of dead in a given period and subtracting average mortality for the same period from that figure. In this case, the demographers and epidemiologists at Inserm chose the period from 1 August, when both heat and mortality began to rise, to 20 August, a week after the heat broke, but when mortality noticeably tapered off. In 2003 that period witnessed 41,621 deaths. The Inserm team compared this to an average mortality of 26,819 for the same period in 2000, 2001, and 2002, and concluded that excess mortality for 1–20 August 2003 was 14,802 deaths. The method "had the advantage of being simple and robust," according to the report's authors, who also acknowledged its shortcomings. As France had undertaken a census in 1999, the researchers were working with a recent assessment of the population, and the report assumed for the sake of expediency a rough constancy of the population between 1999 and 2003. While the authors acknowledged that they could have modeled changes in the overall population (and could have thereby perhaps obtained a slightly more accurate global mortality figure for the period), that modeling would have precluded the careful dissection of their findings by age group that constituted the report's finesse.[27]

The downside of this method is its inclusiveness. It subtracts an average of mortality from total mortality experienced during the period, and portrays the excess as the disaster's toll. It presents the excess heat of 2003 as the killer of 14,802 victims. In broad strokes the method works. In the absence of another major cause of death—for example, a cholera or flu epidemic—the excess heat is the likeliest cause of excess death. But the method also includes deaths that may be only peripherally connected to the heat. During heat waves, people swim in order to cool off. Some may be inexperienced swimmers; drownings therefore usually increase in such periods. But should deaths by drowning count as heat deaths? The heat and its associated sleeplessness may cloud thinking. One subject I interviewed said that during the heat wave she found herself so disoriented that she walked into walls in her apartment. It is not a stretch to think that such disorientation could increase the numbers of accidental deaths or even suicides during heat waves. According to the Inserm model, all of these deaths are part of the global figure.

The differences in the figures each of these models produces can be significant. When another major heat wave hit France in July 2006, the Health Ministry reported that the heat "caused the deaths of 133 persons."[28] Health Minister Xavier Bertrand and InVS director Gilles Brücker touted the health prevention plan implemented in 2004 as a major factor in reducing mortality to such levels. But this study was based on death certificate data, and counted only deaths where heat was listed as an explicit cause of death. The following summer, demographic teams at InVS and Inserm published studies that measured excess mortality, revising the official totals upward to between 1,600 and 2,000—that is, between twelve and fifteen times the original figure.[29] As the InVS team noted in regard to the early estimate of 133 deaths, "This collection of deaths identified as linked to the heat is not exhaustive. . . . Imputing death to the heat entails, in the absence of a case definition, an amount of subjectivity. Moreover, even if a case definition could be used for certain pathological entities directly linked to the heat (dehydration, heat stroke), deaths indirectly linked to the heat (cardiac decompensation, etc.) are often not accounted for."[30]

The inclusiveness of excess mortality measures is somewhat problematic. The median mortality of any given period may not represent a true average, as mortality can fluctuate greatly from year to year. August 2003 might merely be a statistical outlier, with a clustering of deaths by natural causes that accounts for at least some of the excess mortality. But a far more problematic element of such inclusiveness is the concept of "mortality displacement" or the "harvesting effect" of heat waves. The idea behind the harvesting effect is that heat waves (as well as cold snaps and high-pollution events) do not so much cause excess mortality as they *hasten* mortality for those already in poor health. The populations who die during these events are typically fragile ones: they are the elderly, the homeless, and the mentally ill. The harvesting effect concept considers these deaths to be the product of a forward displacement of mortality. If a heat wave witnesses excess mortality, the theory argues, then the weeks or months immediately following the heat wave will experience a concomitant drop in mortality, indicating that the deaths associated with the heat wave represent an advanced, rather than a truly increased mortality.

There are several critical problems with the harvesting effect concept. First, it is extremely difficult to measure. Just as multiple factors can explain elevated mortality levels, they can also explain low mortality levels. High heat might account for significant deaths among the elderly in the summer, but a low mortality level among the elderly the following winter might result less from mortality displacement than from, for example, a mild flu season (which indeed characterized the winter of 2004). Second, there are no clear criteria for how to measure displaced mortality. If I die tomorrow, I cannot die in the future. But when do we consider my death "displaced" rather than "caused"? If it advances my death by a week? A month? A year? Two years?[31]

As a result of these debates over the terms, studies have produced remarkably divergent accounts about whether the 2003 heat wave included a harvesting effect or not, generating significant controversy.[32] One team found a near-complete harvesting effect in the disaster, noting a reduction in expected mortality of some 14,000 deaths in the first six months of 2004.[33] Another study—this one by a climatologist—found a similar "reabsorption" of excess mortality in the low levels experienced in 2004.[34] But most studies found less conclusive results. A systematic analysis of the major literature on the heat wave that appeared in 2005 found that harvesting "only concerned a fraction of the victims."[35] In the short term, the epidemiologist Grégoire Rey indicated, there was no evidence of harvesting, as mortality figures for the month after the heat wave indicated significantly *higher* than expected numbers of deaths.[36] It was only in the very long term—in the period from January 2004 to January 2005—that one could find a significant drop in mortality. But as Laurent Toulemon and Magali Barbieri from the Institut National d'Études Démographiques argued, a dip in mortality itself is not evidence of harvesting. Those studies that noted a reduction in mortality in 2004 did not account for the specific dynamics of flu in the 2003–4 and 2004–5 seasons, characterized by the early arrival of flu in December 2003 (with concomitant mortality) and a relative absence of flu from then until January and February 2005, when mortality again spiked. As Toulemon and Barbieri also noted, such studies did not closely examine the geography of mortality and its absence. If there were a mortality displacement, then one would expect reduced mortality in 2004 in the same areas that experienced high excess mortality in 2003. Yet their data showed that the regions marked by reduced mortality had almost no correlation at all with sites of high mortality the previous summer, a result that signaled the anomalies of the flu season rather than mortality displacement as the explanatory variable.[37]

Another problem with the idea of a harvesting effect operates in an ethical register. The attribution of a death to displacement implies that the death is somehow less real because it advances an inevitable outcome. But all of our deaths are inevitable, so the amount of time lost assumes critical importance. To ascribe deaths to harvesting or displacement therefore devalues certain forms of life and trivializes death. It discounts mortality by indicating that the dead were likely to die soon in any event. By having one foot in the grave, according to this logic, it is as if those at the margins of living had sacrificed the legitimacy of their remaining life.

One team that found significant harvesting in the heat wave noted this point. Alain-Jacques Valleron and Ariane Boumendil, who asserted that there was a long-term harvesting effect that accounted for virtually all of the heat wave's deaths, wrote that "the existence and the size of a harvesting effect is always the object of extremely lively debate: the reference to a harvesting effect is often seen as a means of minimizing the significance of the number of deaths by explaining them as a simple anticipation: those who died were 'destined to die anyway' in the days that followed the event." They also noted that it was important to look beyond blunt measures of data in assessing the heat wave's effects: "It was not only the number but also the circumstances of death in isolation, in overcrowded emergency facilities, etc., that were considered scandalous. Maybe this person was in such a state that his death was already on the way, even without the heat wave. But not that one! This notion escapes epidemiological modeling."[38] The French Senate's investigation of the disaster upheld this notion, proffering what one epidemiologist called a "political" rather than a "scientific" argument against harvesting: whatever the science showed, the report reads, "the lives of these victims were cut short and they died under conditions that are unacceptable for our country."[39]

Another ethical implication of the harvesting effect concept is that those who advocate for its significance often do so to promote a particular agenda. The French Health Ministry offered one example: to consider these deaths as inevitable and due to natural causes would alleviate much of the state's responsibility for mismanaging the crisis. But climate-change deniers are another camp that can find great explanatory value in the concept. Robert Davis, a geographer in Virginia who has published a number of papers refuting anthropogenic climate change, offers a case in point. In a volume published by the George C. Marshall Institute, Davis cites the example of a 1966 heat wave in Philadelphia, in which three days of high mortality were followed by three days of reduced mortality, thereby reducing the net impact of the heat to nearly zero: a clear case of cherry-picking data to suggest that concern about heat waves in a changing climate is overblown.[40] The insurance industry also finds great potential value in understanding mortality displacement. One French actuary studying the problem drew on Toulemon and Barbieri's study—which denied any significant harvesting—to note that the heat wave, in their estimation, entailed a loss of some 100,000 life-years: where Toulemon and Barbieri found that some 4,000 of the heat wave's victims were likely to have died before the end of 2004 in any event, they concluded that the remaining 11,000 would have each lived (at least statistically) for an average of eight to eleven years longer in the absence of the disaster. The actuary concluded that such accounting "opens new perspectives" for "insurers managing life risk on the long term."[41]

Such calculations have the capacity to influence the dehumanization that characterized the treatment of heat victims. To qualify disaster mortality as "displaced" rather than "caused" is to deny its full significance, and therefore the full significance of life lost. The harvesting concept reduces the humanity of the victim by rendering the death less tragic and less substantive than the loss of a death by cause. Agamben's notion of bare life and Butler's elaboration of ungrievable life help to illuminate the epistemological violence that undergirds the notion of a harvesting effect. Both Agamben and Butler signal the capacity of language for modeling differential valuations of life. Agamben's notion of a humanity reduced to the limits of raw existence suggests how such a frame trivializes death by ignoring the significance of life. Such a death is politically invalid and socially unimportant. Butler's concept of ungrievability develops this process. Writing in the context of the war on terror, she argues that the lives of entire populations simply do not "count," that they are not what she calls "grievable" in their death.[42] Butler here points to a phenomenon of dehumanization with complex yet inextricable links to violence. While the lives of American soldiers and the victims of the World Trade Center attacks are grievable, those of not only terrorists but also the civilian victims of American bombing campaigns suffer a double violence: the violence of their deaths, to be sure, but also that of the "derealization" that precludes their grievability. They are anonymous figures at the fringes of life, incapable of assimilation into our imaginative experience. As a consequence of their dehumanization, violence against them "leaves a mark that is no mark," because there is no easy identification with the victim.[43]

The heat wave disaster is a radically different context from the war on terror. But this logic of grievability, derealization, and dehumanization has an important relevance for the ways in which the deaths of the forgotten evoke a rupture between the categories of vulnerability and citizenship. To ascribe a death to displacement is to emphasize a life's insignificance and its unimportance—indeed, its unassimilability to lived experience, and ultimately, its inhumanity. The anecdotal lives of the forgotten evoke no familiarity or even empathy: instead, only a strangeness prompted by a biography framed through its terminal isolation. Their addictions, their disabilities, their social reclusion, effect a violence of derealization that precedes and in some ways preordains their deaths in isolation, for which the heat is less a cause than a catalyst. To speak of their deaths in terms of "harvesting" or "displacement" both highlights and extends the effacement and denominalization that attends both their lives and deaths.

These deaths are ungrievable because they are those of figures whose difference remains unassimilable into the ideals of universal republican citizenship. With few exceptions, the constituent body of the forgotten is drawn from the margins. They were the very elderly poor, the homeless or precariously housed, the addicted, the disabled, and the mentally ill. As Joan Scott has brilliantly argued in the context of the debate over political equality in contemporary France, certain groups have encountered powerful resistance to their claims to full enfranchisement in the republic by virtue of the logic of universalism. Republican universalism in France requires the subsuming of particularity to representability: any citizen must be able to stand for the nation through a divestment of particular or social characteristics. French women and Muslims are two groups who have found their particularities—those of sex on the one hand, and of ethnicity and religion on the other—close to insuperable in the struggle for citizenship by virtue of a sort of unrepresentability that has historically rendered them incapable of standing for the nation.[44]

There is a strong parallel between these cases and those of other unassimilable populations who live at the limits of citizenship, with real effects for the marginalization that this book describes. Here I am concerned less with formal political citizenship than with general social enfranchisement. And it is critical to recognize that for most of the forgotten, their marginalization was an acquired condition. Where the political exclusion of women and Muslims has been based on a perceived ineradicable difference that is incapable of assimilation, the case of the forgotten differs in an important way. For most of the forgotten of the heat wave, social exclusion came as a result of a process of perceived decline or degradation. As I argued in chapter 4, the elderly in particular have experienced a political and social exclusion. The marginalization of the homeless, the mentally ill, the addicted, and the disabled follows a similar rhetorical declension. While Robert Castel has connected the precariousness of citizenship with the apotheosis of the market, there are also deeper roots that mark a broader disaffiliation and exclusion that have distanced some populations from social enfranchisement. As the sociologist Julien Damon has argued, exclusion in the medieval period was not a process, but a punishment that entailed exile from the city, and hence a loss of citizenship. In the contemporary world, exclusion has become less formal: while remaining in place, many in the contemporary city—the homeless, the elderly, the disabled—who have retained their rights of political citizenship describe themselves increasingly as "excluded" from the social body. As a number of scholars have argued, citizenship itself has thus become less a measure of expression at the ballot box, and more about the integration of individuals into the public sphere on a community level.[45] As with the dehumanization of the elderly discussed in chapter 4, descriptions of the deaths of a number of the younger victims of the heat wave call attention to the limitations of social or community citizenship among other marginal populations. The manifestation of anecdotal life that I described in chapter 2—that is, a particular form of biographical storytelling that reduces the significance of life and attributes blame to the victim—assumes a particular salience when the life in question falls outside the acceptable profile of risk, but still ends in a harvested death.

"Nothing to do with the heat": Reaping Death at the Margins

The Rue des Vignoles is a longish street not far from the Père-Lachaise cemetery in Paris's twentieth arrondissement. At its southwestern end, which intersects with the Boulevard de Charonne, it has undergone a process of gentrification in the past decade. Fashionable cafés and restaurants are opening, and real estate prices are going up, although they are still well below the median for Paris and on the low end for the neighborhood. Farther down the block the urban renewal process has been not been as complete, but is in process. Several new housing blocks are under construction. Close to the street's northeastern intersection with the Rue des Orteaux, one is going up in the shadow of a massive and deteriorating apartment building across the Impasse Satan. But in 2007, this construction block was still a row of low-lying buildings. One was a metal shop; there was also a decrepit residential hotel and bar. Several of them were condemned, their windows bricked up, their doors boarded over and padlocked (fig. 24). A number of homeless people had been living in these buildings as squats for several years. On the days I visited, a stream of homeless men and women shuffled between these buildings and the hotel bar's terrace, stopping there to rest and drink. Some carried bags with their belongings; others moved slowly down the street empty-handed, supporting themselves on the barriers that prevented cars from parking on the sidewalk.

FIGURE 24. Squat where Françoise died on 7 August 2003. Photo by the author.

In early August 2003 Françoise was one of the women living in these squats. Françoise had been born in Limoges in 1960. She had been married, but at age forty-three she found herself living in the Rue des Vignoles homeless community. She was an off-and-on heroin user and a heavy drinker, and at some time in the stifling heat of the afternoon of 6 August, she died in a condemned building. Her death notice, dated 7 August at 5:00 p.m., says that the police discovered her about a day after she had died, and that the investigating officer found her "in the common parts of the building."[46]

The death notice listed the address where police found her as 90, rue des Vignoles. This was the metal shop that still stood there in 2007. One of the shop workers I spoke with then knew the community of the squats well. He had been there during the heat wave, and remembered Françoise's death. He told me that she had not died at number 90, but instead next door at number 92. But she had actually lived across the street in number 91, another squat. He also told me that Françoise's death had had nothing to do with the heat—that it was an alcohol overdose. She had been a heroin user and alcoholic for years, but she and the homeless she lived with had been too poor to buy heroin in the months that preceded her death, relying on alcohol exclusively.

During a second conversation with the shop worker, a man emerged from the squat where Françoise had died. He had long, matted hair and a beard and looked as if he had not bathed in days. It was not quite eleven in the morning, and he held an open can of beer in his hand. The shop worker introduced me to him as "Françoise's lover." The man said that she had lived in the street with him for eight years. When I asked about Françoise's death, he reiterated what the shop worker had said: "That didn't have anything to do with the big heat wave." Instead, it was years of drinking, drugs, and homelessness that had deteriorated her health, and an overdose of alcohol that took her life in August 2003.

There is no reason to doubt these accounts of Françoise's death. While neither claimed to witness the event—her body was found about a day after she had died—both men lived near or with her and knew her health status at the time of her death. But the men's description of her death closely matches other accounts of the deaths of younger victims during the heat wave, particularly those with compromised health status. These accounts raise important questions about the inclusiveness of demographic measures of disaster as well as the attribution of death—particularly among marginal populations—in situations of crisis.

Jean-Pierre, for example, lived alone in a single room under the rooftop of a building in the Boulevard Magenta in the tenth arrondissement near a string of adult video stores and bars (fig. 25). According to neighbors, he was

FIGURE 25. Boulevard Magenta, tenth arrondissement, where Jean-Pierre died on 11 August 2003. Photo by the author.

obese—weighing more than 250 pounds—and was a heavy drinker. In one neighbor's account, Jean-Pierre had drunk nothing but beer during the heat wave, rising in the morning and heading straight to the bar across the street from his apartment. On 11 August, a day when the overnight low dipped only to 75°F (23.9°C; at midnight, it remained almost 90°F or 32°C, and certainly some twenty degrees hotter in his sixth-floor attic apartment), the fifty-six-year-old collapsed in the street at midday after spending the morning drinking at the bar. An ambulance took him to the nearby Hôpital Lariboisière, where doctors pronounced him dead at 2:15 p.m.[47] His neighbor told me that Jean-Pierre had told him the heat was unbearable—that he had had trouble breathing since the temperature had spiked—but that it was the drinking and obesity that had ultimately killed him.

Roger's story had some parallels with Jean-Pierre's. At forty-eight, he had a severe drinking problem. A few weeks before the heat wave, a domestic dispute with his partner put him on the street. With few options, he turned to a friend who owned a boardinghouse in the ninth arrondissement. The friend offered him a sixth-floor walk-up room under the building's roof. The room was situated at the end of a long, unlit hallway with bare, unfinished wood floors and cracking plaster walls. The apartment's one window looked out over the courtyard below, but also at the zinc roofs of the surrounding buildings (fig. 26). On the day that I visited the building, the sun blazed on those roofs, reflecting the light and heat directly into the apartment. On

FIGURE 26. View outside the room where Roger was found dead on 10 August 2003. Note the zinc roofs, which radiated heat across the courtyard into the room. Photo by the author.

10 August, when the owner found Roger's body, the exterior temperature was 100°F (37.8°C), and the sky was cloudless. The owner attested to the heat in the apartment, saying that Roger had complained about his sleeplessness. Still, the owner blamed Roger's death primarily on his drinking and not on the heat.

Valérie died under somewhat different circumstances. At only forty-two years old, Valérie lived in a second-floor apartment in a large public housing complex in a quiet residential neighborhood in the nineteenth arrondissement (fig. 27). According to several who knew her, she was an AIDS patient who had entered an advanced stage of the syndrome, and had been quite ill for weeks or even months preceding her death. She had stopped leaving her apartment entirely, but received regular visits from nurses and her mother. She was desperately poor. Both the building's *gardienne* and a neighbor indicated that it was Valérie's affliction and not the heat that caused her death, although they also both indicated that it was some time between her death and the discovery of the body on 20 August. They also suggested to me that since my major concern was vulnerability during the heat wave, I should really be asking about the few elderly residents of the complex who had died.

Finally, Georges represents another dimension of vulnerability. Also a resident of the nineteenth, he lived about two miles from Valérie, not far from the Boulevard Périphérique on a hilltop near the Pré-Saint-Gervais metro stop. The public housing complex in which he lived dates to the early twentieth century, but the wall surrounding it is constructed from the rubble generated by Haussmann's rebuilding of central Paris; the building is somewhat famous as the former residence of the Communist activist Roger Caron. Georges died on 6 August, relatively early in the heat wave, ten days before his sixtieth birthday. According to the building's *gardienne*, his death, too, was a result of his poor health and not of the heat. Georges had an amputated leg, and as a result he rarely left his ground-floor apartment. A heart attack was the probable cause of death in his case, she said.

Informants who knew these figures described marginal subjects who would be at elevated risk for dying at any time. Three were active alcoholics living in precarious conditions with little shelter from the heat. The other two had significant health problems. All except Françoise lived outside of protective social networks, and Françoise's situation was anything but ideal. In each case informants suggested that the heat was at most only an accessory to their deaths—that the real cause of death was a long-standing health problem, and the heat merely pushed them over the edge. Yet all of them were buried at Thiais in August and September 2003, and all of them figure in the mortality calculations of the heat wave. Valérie even became something of an iconic figure of abandonment during the heat wave: her body was the first interred during the 3 September ceremony at Thiais, and one of only two interred in the presence of President Chirac and Mayor Delanoë.[48]

These cases point in particular to the difficulty of measuring mortality in

FIGURE 27. Public housing project, nineteenth arrondissement, where Valérie was found dead on 20 August 2003. Photo by the author.

a period of disaster. As individual cases it would be easy to write them off as the types of subjects who find their end at Thiais in normal circumstances. And the unreliability of memory itself could be a critical factor in explaining why it was so easy to ascribe the deaths of some to the heat and those of others to natural causes. A 2006 forum in the *American Sociological Review* on the Chicago heat wave of 1995 pointed to the acute nature of this problem. As Eric Klinenberg argued in the forum, fieldwork conducted long after a disaster has passed faces a number of challenges. Not only are memories often cloudy after the passage of several years or more—he cited informants who had no recollection of deaths that had occurred in their buildings or in adjacent properties—but it can also be difficult to find sources with direct knowledge of a given death, leading to a reliance on hearsay. Finally, considering the media's and public officials' widespread blaming of families and neighbors of those who died, it is not surprising that the stories they tell about the victims might have an air of defensiveness about them.[49]

Similar factors were at play in my interviews surrounding the forgotten victims of the 2003 heat wave. It is easy to see how several years could cloud memories of the disaster, as well as how the politics of blame could foster a gradual retrospective revision of the events that took place, prompting those who surrounded the victims to see the victims' personal failings and already compromised health as the principal factors involved in their deaths, rather than something that their interventions could have easily prevented. Yet it is clear that such memories of the dead constitute an important form of evidence in their own right. As I argued in chapter 2, the narration of these lives constitutes an important interpretive frame. Although the reliability of these informants' claims remains open to question, it sheds significant light on the retrospective construction of the experience in a broader memory of the disaster. What is significant is not necessarily whether one memory is more accurate than another; instead, it is the near-complete uniformity of storytelling about the victims and their lives that provides a context for understanding the cultural meaning of the disaster and death at the margins of society.

As a group these particular victims and other similar cases raise questions about who is a casualty of disaster. They were all marginal figures: no one would be surprised to hear of the death of a homeless alcoholic and sometime injecting drug user, an alcoholic living alone in an attic apartment, a late-stage AIDS patient, or an isolated and disabled man in poor general health. Informants who knew the victims signaled the heat as at most a contributing factor rather than a cause of death, some even pointing to the elderly as the real victims. The thrust of most of these conversations was that these subjects would have died regardless of the weather. Perhaps the heat pushed them into the grave a bit early, but it was a track along which they were rapidly headed in any case. Yet the heat could well have significantly accelerated any of their deaths. Dehydration would set in much faster in alcoholics than in the sober, heat stroke would quickly capitalize on an AIDS patient fighting opportunistic infections, and cardiac problems are a major attendant cause of death in heat waves. Possibly most significant is the reflection of excess mortality at Thiais. The cemetery's *secteur d'indigents* normally receives thirty bodies in a given August. In August 2003, it received over a hundred. Probability suggests that although some of these deaths were indeed a result of factors other than the heat, most of them were heat related.[50]

The notion of a harvesting effect is so persuasive because heat strikes the already vulnerable. As with the very elderly, people with certain addictions and disabilities live at heightened risk for death from heat stroke. But they also live at a constant state of high mortality risk from a range of other circumstances. Influenza that might confine a forty-year-old to bed for a week can be deadly to an eighty-five-year-old. Addicts are far likelier to overdose than the general public. The mentally ill are at elevated risk for suicide. And violence and deadly infectious diseases are factors of everyday life for the homeless. If the heat is a reaper of mortality rather than its cause because of the fragility of these victims, are these other causes of death in vulnerable populations not also harvesters in their own right, if more evenly distributed ones?

In the accounts of their neighbors, Françoise, Valérie, Roger, Jean-Pierre, and Georges were collateral damage of a disaster rather than its true victims. According to this logic their deaths become less significant in the gross calculation of mortality. The attribution of their deaths to addiction or to poor general health suggests that they constitute part of the baseline mortality—the median figure in any given year—and not part of the excess mortality that made the heat wave such an egregious disaster. And yet they are exactly the types who are the most vulnerable to death during a heat wave. Given their status as inhabitants of precarious life, would they have died in the next few weeks? Months? Years? The particularly frail, demographers suggest, are likelier to die in the earlier stages of a heat wave.[51] But of these five, only two died in the first week. They are the types of subject who make a compelling intuitive case for the idea of mortality displacement or a harvesting effect. Yet the aggregation of so many deaths in similar circumstances indicates that the harvesting concept is more exculpatory than explanatory, amounting to a dismissal of life on the margins rather than a clear picture of disaster's effects on mortality.

Aggregation and Representation

The cases above indicate the murkiness of disaster epidemiology. By official cause of death, not one is a heat victim. Measured by excess mortality, any of them might be. So would Minh, a sixty-six-year-old immigrant from Vietnam who killed himself in his sixth-floor apartment at the peak of the heat wave; Daniel, a forty-six-year-old homeless man who died on the doorstep of a shelter in the eighteenth arrondissement; or Noël, a forty-eight-year-old whom paramedics found in a park in the nineteenth arrondissement "dressed as if for winter" in the 100° heat before he died in the Hôpital Lariboisière shortly thereafter.[52] The heat might have been implicated in any of their deaths, or in the deaths by drowning that also went up by nearly 40 percent during the summer of 2003.[53] But where a tally by death certificate data is too exclusive to provide a realistic picture of the disaster's toll, a measurement of excess death is too blunt an instrument to distinguish anything but aggregate patterns in mortality.[54]

Such aggregation is a useful tool. As Denis Hémon and Eric Jougla argued in their Inserm report, global measures of excess mortality allowed them to compartmentalize and to compare mortality by age, sex, and region with a degree of sophistication that would have been unfeasible with other methods.[55] They were thus able to indicate the extremely elevated mortality of those over seventy-five (75 percent higher than normal) and the much smaller effect for those under forty-five (8 percent higher than normal). The study indicated that mortality more than doubled for those over ninety-five.[56] It also revealed that mortality in the Paris region was nearly two and a half times the normal rate, but that it was only 20 to 30 percent higher in the South of France (Languedoc-Roussillon and Provence-Alpes-Côte-d'Azur)—where mean temperatures were higher than in Paris—thus indicating that mortality was not linked to absolute heat load.[57] The aggregate picture of risk thus showed the elderly as the principal vulnerable population in the heat wave. Over 80 percent of those who died were among the very elderly. Elderly women in particular bore the brunt of death: women over seventy-five constituted nearly 60 percent of the global mortality figure.[58]

This model of vulnerability is a useful tool for allowing the state and local communities to prioritize their resources. There is a predictive component to the report: an assumption that heat waves are particularly deadly to the elderly for physiological and social reasons. Policy initiatives in the aftermath of the heat wave have aimed decidedly at the elderly as a target for intervention. In February 2004 the government agency charged with the care of the elderly issued a directive that all retirement homes be equipped with at least one air-conditioned room, for example.[59] Heat awareness campaigns have both implicitly and explicitly signaled the elderly as potential victims. Posters distributed throughout Paris in the summers since 2003 encourage solidarity with the elderly during heat waves, showing photographs of children offering their elderly neighbors bottles of water or aiming fans in their direction. And as chapter 4 indicates, the media emphasized aging as the major contributing factor to elevated mortality almost from the moment they became aware of it, with both television and print journalism dedicating significant attention to demographics and their health implications in France.

The report contains an important implication for gauging the social components of heat risk. The report's data reinforce the importance of the social environment as a critical factor in shaping vulnerability. Although the report points to the elderly as the group that suffered the highest excess mortality in France, the data also suggest extremely low excess mortality for another group: small children. Newborns and infants are extremely physiologically vulnerable to high heat: they lack adults' capacity to dissipate heat through sweating and have a low surface area-to-volume ratio.[60] And yet according to Hémon and Jougla, only a handful (six) more children under age one died during the heat wave than in normal circumstances. In a special issue of the *Bulletin épidémiologique hebdomadaire* dedicated to the heat wave, Hémon, Jougla, and several coauthors went even further, asserting that "small children were not an issue" for elevated mortality.[61] A report by InVS echoed this claim, noting that "very few pediatric deaths were observed," despite the fact that "very young children also constitute, *a priori*, subjects at risk during a heat wave."[62]

The observation about infants' resilience during the heat wave is critical, because it highlights the importance of social rather than physiological risk factors in determining outcomes. Infants are physiologically as vulnerable as the elderly for heat illness and death, but they have social protections that extend far beyond what most elderly enjoy. Infants are born into networks of care, either in the family or as wards of the state. Parents recognize their essential vulnerability and care for their basic needs: feeding, hydration, clothing, and shelter. Of course some children suffer extreme neglect.[63] But in general, parents recognize infants' suffering and respond to it quickly: as the InVS report notes, better "parent education" as well as closer "medical surveillance at this age" has "considerably reduced child mortality" during heat waves.[64] And the community responds rapidly to vulnerable infants as well. One sociologist who lives near a hospital in Paris described a moment of solidarity with newborns during the heat wave, telling me of café owners eagerly bringing ice to the hospital "for the babies."[65]

In contrast to infants, many of France's elderly have aged out of their social networks. Their vulnerability is physiological, but also social. This is an important limitation of assessing risk through the tabulation of death. The elderly constituted the bulk of the mortality burden during the heat wave for a range of factors. But they are perhaps most significant as the largest group of social isolates in France, with high poverty levels. Their age and economic precarity have made them vulnerable to isolation, which in turn structures their vulnerability to extreme heat. And if it is a social ecology of isolation rather than (or even in addition to) physiological vulnerability that elevates the elderly's risk of dying from heat illness—which the high survival of newborns and elderly in socially integrated and high-care situations suggests—then other isolated, marginal, and poor populations are also at elevated risk during extreme events. As other studies have confirmed, it was not just old age that structured vulnerability during the heat wave. The sixteenth arrondissement has the highest concentration of elderly of any administrative unit in Paris, but had among lowest rates of excess mortality during the heat wave; it also has among the highest household income levels in the city.[66] Given similar heat loads, poorer neighborhoods throughout the country bore a higher mortality burden than did rich ones.[67] As another study put it, it was the elderly poor—rather than the elderly in general—who died in such high numbers.[68] And paradoxically, it was the "frailest" of the elderly who survived the heat wave in the greatest numbers. Many who died lived in retirement homes. But the institutionalized population in general did better than those living at home (19 and 35 percent of excess deaths, respectively), and among those who died in nursing homes, it was the bedridden patients rather than the ambulatory ones who proved to have the greatest resilience, precisely because staff paid closer attention to the former than to the latter.[69]

Because the "typical" heat wave victim, according to epidemiological studies, was elderly, age has become the principal narrative of the heat wave. Yet it is only one of many stories of risk and vulnerability during the disaster. One cannot deny the significance of the death toll among the elderly, but a focus on old age per se—even one that attends to the social components of aging as well as the physiological ones—misses a crucial element of the catastrophe. Hémon and Jougla urge those reading their report to consider a range of social factors implicated in heat death, including "characteristics of the urban micro-environment as well as those of living spaces, individuals' economic and social environment, their health status and their access to medical care, and their medico-social environment."[70] As investigations of other heat waves have shown, factors such as addiction and mental illness are as important as age in determining risk for heat illness and death.[71] It is less old age than a combination of social marginalization and physiological vulnerability—owing to poverty, old age, dependency, disability, infectious disease, addiction—that ranks among the most significant risk factors for dying during a heat wave. Poor housing, poverty, compromised health status, and inadequate access to care are epidemic among France's elderly, but they are not unique to them.

The stories of Françoise, Roger, Jean-Pierre, and others are limited in what they can tell us. Like all the narratives of the forgotten, they are fragmentary and provide only glimpses into their experiences of life and death. But as with Suzanne, the case with which I began this chapter, they offer insight into a vernacular understanding of the impact of the heat wave and risk in the urban environment. The ease with which neighbors mistakenly associated the elderly Suzanne's death with the heat wave suggests the extent to which the idea of the elderly as nearly the exclusive victims of the disaster has taken hold. So, too, does the resistance of those who knew Françoise, Roger, Jean-Pierre, and the others profiled in this chapter to the idea that the heat might have played a significant role in their deaths.

Another case of risk profiling offers a useful example of the inherent dangers of such a practice. In the summer of 1981, a handful of men with histories of multiple sexual relationships with other men and drug use showed up in hospitals in Los Angeles, New York, and San Francisco. Doctors watched helplessly as the patients' organ systems failed one by one as a result of opportunistic infections—thrush, pneumocystis carinii pneumonia—that were normally either harmless or typically seen in infants, the elderly, or the immune-compromised. When the physicians published their observations in the Centers for Disease Control and Prevention's *Morbidity and Mortality Weekly Report*, they emphasized the patients' sexuality as closely linked to the new syndrome. An editorial note in the second of these articles alerted physicians to pay attention to their patients' sexuality: "Physicians should be alert for Kaposi's sarcoma, PC pneumonia, and other opportunistic infections associated with immunosuppression in homosexual men."[72] A wave of subsequent studies—each relying on the aggregation of cases—assigned a central role to sexuality in a rapidly spreading epidemic, which authorities at the CDC quickly labeled "Gay-Related Immunodeficiency," or GRID. Although such studies employed solid data—indeed, most such cases *did* occur in men who had sex with men—a consequence of this research was a widespread ignoring of other groups at risk for acquiring the syndrome, including injecting drug users, sex workers, and populations such as hemophiliacs who regularly received blood products. The rapid accumulation of outlier cases quickly forced a reconsideration of the near-complete association of the illness with male homosexuality—and its renaming as AIDS, or Acquired Immunodeficiency Syndrome—but the profile remained seared into both medical and public consciousness. As a result, a growing AIDS epidemic among women went largely unnoticed through the 1980s, and a vehement denial of the possibility of heterosexual AIDS persists in the present, with grave consequences for policy and practice in affected areas.[73]

The modeling of vulnerability is among epidemiology's most useful tools. But it brings risks of its own: chiefly, as it develops knowledge prodigiously in one domain, it tends to obscure others. In the case of AIDS, an emphasis on the importance of homosexuality to the disease's transmission concealed a growing heterosexual epidemic in women and developing countries, delaying appropriate interventions, with deadly effects. In the case of the heat wave, an emphasis on the elderly to the near exclusion of all other vulnerable groups obscures the role of poverty and isolation as critical risk factors at all levels, and not just among the elderly. The narratives that informants shared with me suggest productive ways in which the anecdotal can complement the aggregate in modeling vulnerability to disaster, and can signal populations at risk. Like the notion of a harvesting effect, ignoring the potential role that a disaster plays in killing the poor and the unhealthy mitigates disaster by denying its force. To attribute their deaths to their physical fragility rather than to extreme weather or to a decaying social ecology is in a way to erase the significance of their deaths, and therefore of their lives. Their experiences on the margins point to a critical relationship between health and social citizenship—a tendency according to which degrees of political, social, and economic enfranchisement serve as important determinants of health. A focus on the vulnerability of elderly populations is essential, but it risks overlooking some 3,000 other victims who died during the crisis, and the role that state and community can play in extending its protections to them.

Epilogue

Thierry Jonquet's marvelous novel *Mon vieux*, which is set in the summer of 2003, poses an ethical debate for its readership about the value of life and provides a stark illumination of how the heat wave preyed on social inequalities.[1] Alain Colmont is a television screenwriter of middling success. Divorced and fifty, his greatest writing is behind him, and although he has a bit of money saved, he is worried about making ends meet in the future. His daughter, Cécile—now nineteen—suffered a motor scooter accident three years earlier that left her horribly disfigured. After multiple reconstructive surgeries, she has regained most physiological function, but she remains unrecognizable. Since the accident, she has suffered from profound depression, and Alain is convinced that with extensive plastic surgery she can recover completely and live a full life. An excellent surgeon in Paris is willing to take on the challenge out of professional interest, but will charge 60,000 euros (about Alain's annual income) for the procedures. The state refuses to pay for the surgery, insisting that although Cécile's appearance is grotesque, there is nothing medically wrong with her. Alain is willing to risk financial ruin to give his daughter a chance at life. Yet about a third of the way into the novel, Alain's father, Mathieu, reenters his life. Mathieu had abandoned his wife and son over forty years earlier, leaving them destitute. Now, Mathieu appears in a hospital suffering profoundly from Alzheimer's disease. The seventy-five-year-old does not recognize his son, nor does he express any regret at having abandoned him. He is obstreperous with the hospital staff, rejecting his medication and uttering little more than racial slurs at them. Most important for Alain, the state insists that he is liable for Mathieu's medical care: he must pay roughly 60,000 euros in back hospital fees and assume responsibility for his future treatment at great expense. This legal obligation to provide for his father—the figure who abandoned him and who is now a human shell rather than a caring being—means that he will not be able to pay for his daughter's surgery, precisely at the moment she is showing strong signs of recovery.

The arrival of the heat wave in August presents Alain with a dilemma. Alain's neighbor Jacquot (who is well aware of Alain's troubles) works as a switchboard operator at a hospital not far from their apartment building in Paris's Belleville neighborhood. Jacquot returns home from work on 7 August, shocked and exhausted, and appreciative as Alain hands him a cold beer. "Goddamn, you're not going to believe this shit!" he shouts after having drunk a few swallows. "At the hospital, it's madness! But I mean beyond madness!" When Alain presses him to be clearer, Jacquot says: "The little old people . . . it's crazy. They're coming into the ER non-stop, in awful shape. They're dying by the hundreds. . . . We're cramming them into the hallways, wherever we can, but that's not enough! We don't know where to put them anymore!" But then Jacquot quickly shifts the tone of the conversation: "With a little luck, your old man will pass and that'll be the end of your problems."[2]

Within a day or two, Jacquot suggests they not wait for luck to play its part. He proposes that Alain murder Mathieu in order to allow Cécile to flourish. When Alain refuses to consider the idea, Jacquot offers to do it himself. For 10,000 euros, he would commit the murder by sneaking into the hospital and giving Mathieu an overdose of a neuroleptic; combined with his age and infirmity, the drug would surely kill him. With so many elderly dead in the hospital, no one would notice. For days, Alain rejects the idea. Little by little, he starts to wear down. He goes to the hospital himself, armed with an overdose of sleeping medication, but cannot bring himself to murder his father. Finally he agrees to Jacquot's proposition. But Jacquot botches the job—leaving Mathieu alive—and flees the city. Meanwhile a homeless man who has overheard Jacquot and Alain's conversations blackmails Alain. The beleaguered Alain pays the homeless man, who is himself mugged and stabbed to death by another homeless man. At the end of the novel, Mathieu dies of a heart attack a few weeks later, absolving Alain of his legal responsibility. Alain is out the blackmail money, but has enough savings remaining to allow Cécile to begin her surgeries.

As I noted in chapter 4, other novels have used the heat wave to make rhetorical points. But *Mon vieux* goes far beyond Gavalda's or Houellebecq's uses of the disaster. Where Gavalda and Houellebecq each signal the ways in which the disaster revealed a dehumanization of the old that had in their views become commonplace in French society, Jonquet sets the novel itself during the heat wave. His characters sweat. The heat saps their energy. They quench their thirst. Jonquet employs an ethnographic eye in describing the relations among the social networks his book describes: between an unemployed man and the wealthy landlady who allows him to rent a *chambre de bonne* off the books, then pushes him into homelessness when she needs the room for a relative; among the band of homeless schemers who populate the landscape of Belleville around Alain's apartment, who struggle against city authorities as well as within their own group's rigid social hierarchy; between Alain and his neighbor Jacquot; and between Alain and the faceless bureaucrats of a welfare state. The heat operates less as atmosphere than as a character or circumstance that drives the plot forward by complicating the social networks that populate the book. Jonquet tells his readers not only about the sweltering days and nights, but also about difficulties of coping with the heat: the homeless, unable to find space in shelters, retreating to the "entirely relative cool" of the metro platform, sleeping shirtless and waking constantly from the vermin crawling on their bare skin; the middle classes finding long waits at public pools so crowded with bodies as to offer no relief; the dozens of bodies piling up in hospitals.[3] He interweaves the novel's narrative with the media's, quoting television news and offering a day-by-day chronicle of the mounting catastrophe.

Jonquet also raises troubling questions about the value of life. Where Houellebecq castigates (or possibly celebrates) his nation's disdain for the old, Jonquet problematizes it. He confronts his readers with the question Beauvoir posed in *La vieillesse*: should a society invest in its youth, or rescue its old from misery? As Jonquet presents the problem, it appears simple: Alain's father is a reprehensible character who gave him nothing; his daughter is full of promise and loves her father unconditionally. As Danièle Alet pointed out with several figures in her film *Aux oubliés de la canicule*, it was not Alain who abandoned Mathieu to die, but rather Mathieu who had abandoned his family decades earlier. It is easy for the reader to empathize with Alain's dilemma. When he looks forward to his father's death, we understand that it is with relief, not with malice, and that it is only to allow him to support his daughter rather than a repellent, abandoning, and incurably demented man who has shown only selfishness. Yet the dilemma is wrenching for Alain. And although Jonquet sides with Alain and Cécile over Mathieu, the relief when Alain cannot complete the murder is palpable: at this moment, and at the moment when he tries at the last minute to dissuade Jacquot from killing Mathieu, the reader is frustrated with Alain's situation but reassured of his humanity. Mathieu's eventual death brings the novel to a fortunate conclusion for Alain, but it is one that depends (as does the entire novel) on chance rather than intention. Although we sympathize with Alain's situation and eagerly await his father's death, Jonquet also helps us see that a devaluation of elderly life to the point of eradicating it in favor of youth marks a monstrous turn.

A discussion of Jonquet's novel marks a fitting end to this book, not only because of the novel's emphasis on the difficulties of valuing life but also in its unflinching look at the systemic problems of the contemporary city. As with Gavalda's *Ensemble, c'est tout*, the book has a redemptive theme. Yet it has none of the saccharine qualities that one finds in Gavalda. Mathieu does not die anything like a "good" death; instead, redemption comes in Alain's inability to kill his own father despite the overwhelming temptation—and easy justification—for doing so. In addition, Jonquet's Belleville reveals the fracturing of the social contract through his depictions of the homeless bound to misery and degradation by their addictions and sicknesses, destined for misfortune; of the inadequate state institutions that cannot help but fail to get the homeless and the unemployed back on their feet; of the cruelty of both formal and informal market forces that make housing a luxury rather than a basic human right; of a social welfare system whose limitations place any middle-class citizen an accident away from ruin. The heat wave operates for Jonquet as an important plot device, but it operates in the novel as it did in France itself: as a catalyst rather than a root cause of the catastrophe. Instead, for Jonquet, the root causes of devastation are part of the social fabric, the very precariousness that is built into early twenty-first-century French society.

Mon vieux provides a novelistic insight into precisely the problems that this book has attempted to engage. The state is an important character in the novel, but in its abstraction of all problems to general rules it fails to recognize the particularity of individuals. Instead, the novelist's gift for storytelling illuminates the tensions between informal street-level practice and the totalizing and inhumane operation of the state. It is the individual stories of the novel's characters—the anecdotal lives of Alain, Cécile, Mathieu, Jacquot, and the band of homeless who surround them—that tell us far more about the experience of the heat wave than any public hearing or any epidemiological report. The stories of how one becomes homeless while under the protections of a welfare state, of how medical expenses remain a barrier to care even in a generous health system, and of the very personal nature of each case of abandonment and forgetting suggest the importance of listening to such voices in accounting for the toll of disaster.

It is perhaps a failure to account for these voices and experiences that has led the state to do little to address the root causes of disaster in the aftermath of the heat wave. There is a widespread recognition that the disaster was less a public health catastrophe than a social one: even Jean-François Mattei has stated this flatly, both in our 2011 conversation and in repeated interviews with the media. Yet state and municipal authorities have focused on simple solutions to a complex disaster. Drawing on aggregative reporting that points to the elderly as the single largest vulnerable group to heat distress, policy has focused almost exclusively on them. But it has done so in simplistic ways that have done little to address the broader pathologies of poverty, inequality, and social isolation. The state has mandated the installation of air-conditioning in nursing homes. Health authorities now coordinate closely with weather authorities, and cities have established heat plans to cope with extreme temperatures. There are local, regional, and national warning systems that now signal the alert when the temperature rises, and public advertising campaigns that emphasize the importance of solidarity with the vulnerable (see fig. 28). Paris municipal authorities have set up a telephone network designed to operate as a safety measure in periods of high heat: after registering with the city, those who worry for their health during heat waves are placed on a call list. When the temperature rises beyond a certain level, social workers call to ensure their safety. The network represents an important step toward overcoming the problems of broad social isolation: those without family to check on their conditions can rely on the state to do so. But it is not without its problems: according to one anecdote, an elderly woman returned home from grocery shopping to find the firemen breaking down her door, as she had not picked up the phone when authorities had called her.[4]

The city's telephone network also fails to help those who most need intervention. It relies on citizens to have telephones, for one thing. But it also relies on them not only to recognize their vulnerability, but also to avow that vulnerability publicly. Posters advertising the network (including one I saw on the door of one victim's building in the nineteenth arrondissement; see fig. 29) encourage the participation of those who are "elderly," "disabled," or those who feel otherwise "incapable of tolerating summer in Paris (poor health, solitude)" in the program. To register with the city as one at risk during a heat wave requires an admission of solitude and vulnerability, as well as that an individual identify with those who died during the 2003 crisis. Yet the characterization of those who died implicitly emphasized their nonrepresentability, precluding easy identification for the potentially vulnerable. For the poorest citizens, for those with mental illnesses, for those with addictions that overwhelm their lives, self-registration in a municipal network is unrealistic. For those who do not suffer from such conditions, self-identification as vulnerable is unlikely. And as Marie France's case indicates (see chapter 2), even for those who recognize their vulnerability, the admission that their independence is better characterized as isolation is improbable.

FIGURE 28. Electronic billboard in Paris, counseling citizens to hydrate regularly because of the heat, June 2011. Photo by the author.

FIGURE 29. Poster advertising Paris's heat emergency network, identifying the elderly, the disabled, and the isolated as primary risk groups. Nineteenth arrondissement, June 2007. Photo by the author.

Beyond the establishment of heat plans and surveillance networks, there has been some effort to transform a basic landscape of vulnerability in Paris. Mayor Bertrand Delanoë has demonstrated a commitment to making Paris a "sustainable city": one that embodies an ecological sensibility in reducing its environmental impact and one that can adapt to a changing climate. Guidelines for new construction indicate an awareness of the relationship between the built environment and heat hazards, noting the importance of "struggling against the heat-island effect" and "managing risks," including the protection of those "most vulnerable to high heat" (still relying, however, on "voluntary registration").[5] One means of accomplishing this is the use of vegetation on exterior walls and roofs of buildings to operate as heat sinks: even a small proportion of rooftops covered with vegetation can reduce temperatures by one to two degrees Celsius, according to city officials.

There is evidence of these efforts in some of the most marginal districts in which the forgotten lived and died. When I walked down the Rue des Vignoles in June 2012, for example, where Françoise had died in a ramshackle squat, the area showed signs of slow transformation. The squat itself is under construction as a public housing unit, and three more buildings are undergoing similar retrofits on the street. Most of these are traditional buildings, but one new public housing block has embraced many of the ideas the city's development office is promoting as sustainable and protective. The project, designed by the architect Édouard François, is called "Eden Bio." With its multilevel row-house feel and pitched roofs, as well as the massive wooden trellis that surrounds the project—the architect's idea is that wisteria and other plants will soon surround the buildings—it stands in stark contrast to its surroundings, as well as the standard image of public housing, and as such has caught the attention of the green architecture community as a visionary project.[6]

Sustainable development policies represent an important step, but they are insufficient in addressing the broad structural vulnerabilities that exacerbated risk during the heat wave. For one thing, they guide new construction but leave the majority of the urban landscape untouched. When I spoke with Patrick Pelloux, the former head of the emergency physicians' union, he was dismayed at how little had changed since the heat wave, and cited the city's nineteenth-century architecture as a principal culprit. We met in the offices of the satirical weekly *Charlie Hebdo* in 2009, where Pelloux penned a column, and which were then located in the Rue de Turbigo on the border of the second and third arrondissements in central Paris. As we talked, he gestured to a Haussmannian building out the window. "One thing France has not done is the renovation of housing," he noted. "It's very pretty, but Paris is becoming more and more like Disneyland," he said, a fact that city planners were reluctant to change. The fundamental architecture—with its vast French windows and zinc roofs that allow the sun to bake apartments in summer and do little to conserve heat in the winter—dramatically exacerbated risk. "The roofs of Paris are ovens," he said. But that, he insisted, is what the tourists want to see. It was this deadly charm—along with its hidden populations in its upper corridors—that made and continue to make the city so vulnerable, he insisted.

Pelloux also lamented how the health system remained completely inadequate for coping with any new crisis. Resources given to hospitals in the aftermath of the heat wave had been rescinded after the financial crisis of 2008. "Last winter, we had 6,000 people die from a lack of access to care. Politicians said they died from the cold. That is false. Cold doesn't kill. The problem is a lack of access to care." More important, he noted, "We haven't done anything in relation to environmental crises at all." By failing to attack climate change and relying on technological fixes, the state had betrayed the victims of the heat wave; "Now it's six years later. What have they done? Ah, not much. To say, 'We're going to air-condition everything,' that's crazy. That's stupid. That's completely mad. Because that will entail overconsumption of energy, and that overconsumption of energy will have climate change as a consequence." Moreover, Pelloux insisted, the economic crisis after 2008 would have real consequences for increasing social vulnerability in France: "And we will feel this when there is another heat wave."

When I asked Pelloux if he would say the same thing were he to write his book on the heat wave, *Urgentiste*, again, his response was simple: "Exactly." The medical system was "not at all" prepared for a renewed crisis. The 2006 heat wave had brought lower mortality, but that was because its circumstances were different. It was longer in duration but lower in intensity, offering bodies respite and recuperation at night. Therefore it did not constitute a real test of the measures France had put in place after 2003. The one difference, Pelloux asserted, is that after 2003, "the culture, the civilization, the people, what's changed is that now they know that heat kills. Perhaps that's what's changed."

On the other side of the medico-political spectrum, the former health minister Jean-François Mattei is just as insistent in reasserting his views and his positions from 2003. He told me in 2011 that his ministry's alleged mismanagement of the crisis was a media fabrication. "The heat wave was not particular to France," he said, "but the media's construction of it was." Journalists were eager to blow the disaster into a political scandal, but none followed the story once the dust had settled: none noted after the fact that Italy had had a higher death rate, and that other European countries had suffered higher mortality when measured as a proportion of the population. Mattei blamed Patrick Pelloux for propagating the notion that there was a rift between the ministry and the hospitals during the crisis. He also insisted that France's failure to anticipate the health crisis was only normal. He had spoken with American public health specialists, he said, and "they told me quite clearly, each time a heat wave strikes for the first time in a given country or region, there is a catastrophe, because no one is prepared for it, and no one has a measure of the danger of these accidents." And even though the ministry issued health advisories as the heat wave struck, no one took them seriously, "because they didn't seem like health measures, but instead like common-sense measures." Where people took pandemic preparedness quite seriously, they ignored heat advisories, because, he insisted, even among professionals, no one considered heat lethal. In contrast with Pelloux, who insisted that the state's responsibility for the crisis was "total," Mattei argued that the state held no responsibility for the crisis, because the major problem was a social one—the isolation of the elderly—rather than an epidemiological one.

Mattei expressed little doubt about his handling of the crisis, stating that the factors that placed the victims of the heat wave at risk are beyond the Health Ministry's responsibility. Perhaps this is true: forging stronger social connections between the elderly, the mentally ill, the homeless, and the broader population is not the agency's explicit charge. The breakdown of community in the twenty-first century is a social rather than a medical problem. But in a broader sense, as Pelloux argued to me, the heat wave was and remains a health crisis par excellence. The heat wave brought to light "exactly the World Health Organization's definition of health, a state of medical, psychological, and social well-being," Pelloux insisted. If a crisis preyed on those who found themselves "in the greatest social difficulty," he argued, then it is the responsibility of the state to ensure the protection of that population.

The ongoing argument between Pelloux and Mattei indicates that little has come from the political debate that followed the catastrophe. The state remains focused on a utilitarian conception of vulnerability that emerges from an aggregate demographic portrait: if the elderly constituted the bulk of the victims, then attention to their vulnerabilities will mitigate future heat disasters. As one French epidemiologist wrote to me in 2013, "Many (generally those who work far from the field) are very optimistic today and think that the measures put in place since 2003 will allow us to avoid any problems." These same actors see the reduced mortality of the 2006 heat wave "as a success, even though there were still 2,000 deaths." And in subsequent summers, heat deaths have continued, both among the very elderly poor and among the homeless who remain outside the protections of new measures designed to manage disaster in advance.

In an article about the atrocities that have marked the Congo Free State and the contemporary Democratic Republic of the Congo, the historian and ethnographer Nancy Rose Hunt argues for the need to bring new modes of listening to such histories. She argues that historians must "parse the archive" by pushing beyond dominant narratives in these histories, and must instead "seek weaker, more fragile acoustic traces" that can inform us about the relationship between past and present.[7] This is something like what I have attempted to accomplish here. For all the value of official and media inquiries into the disaster—which bring the full armature of state authority, teams of well-trained investigators, and the power of a relentless media spotlight—there remains much to be learned by listening to these weaker traces of the disaster. Just as a novel's depiction of the miseries and indignities faced by a group of homeless trying to find a safe place to sleep out of the heat contains a poignancy that no official inquiry can capture, there is a value to walking in the steps of some of the heat wave's loneliest victims: climbing their stairs, smelling and hearing the mice in their hallways, and above all, listening to those who surrounded them. The stories these informants choose to tell and the tone in which they relate them provide a means of situating disaster in the lives of those whom it struck and the histories that produced them. More than merely revealing inaccuracies and oversimplifications in the bureaucratic knowledge that the heat wave produced, these stories indicate the ways in which official knowledge has perpetuated the marginalization of the victims and facilitated their forgetting.

Early in the summer of 2012, I returned to the Thiais cemetery for a final visit. I was curious about whether any of the forgotten would have remained there nearly a decade after the catastrophe that took their lives. Although official policy calls for the removal of those buried at public expense after five years, most of them were still there at my last visit to the cemetery in 2009, then six years after the heat wave. When I entered divisions 57 and 58 of the cemetery, I saw that most of the occupants had died quite recently, and imagined that the forgotten were now truly consigned to the past. But after searching more closely, I found that nearly seventy of those who had died during the heat wave remained at Thiais. Pedro, Paulette, Marie, Jean-Louis, Roger, Mauricette, and most of the others still occupy the graves in which the republic buried them.[8] Some have been moved to make room for new occupants, but the majority of the victims who populate this book—and the social imagination of the disaster—still lie in the public domain. These bodies are physical yet evanescent traces of a catastrophe that signaled the vulnerabilities not only of France's most marginal subjects, but of the welfare state itself. The disaster derived its strength from social fault lines that have been at the center of political and social debate for decades: about the sharing of responsibilities among individuals, communities, and the state; about the demographic transformation of society; about the renewal of the built environment; and about the meaning, limitations, and protections of citizenship in the contemporary era. Like the voices of Hunt's actors in the Congo, and like the characters of Jonquet's novel, the stories of these bodies and their witnesses contain historical insights that reach far beyond the contents of the state's burgeoning archive.

Acknowledgments

Many people and institutions made this project possible. I am indebted first to those whose financial support allowed me to complete the research. A Science and Society Scholars grant from the National Science Foundation allowed me the time to indulge in extensive fieldwork in Paris, as did a Bourse pour l'accueil des chercheurs étrangers from the city of Paris. A MiRe-DREES grant from the French Ministry of Health offered essential funds to get this project started and to build connections with researchers in France and the United Kingdom. In Paris, the École des Hautes Études en Sciences Sociales and the Institut de Recherche Interdisciplinaire sur les Enjeux Sociaux (IRIS) provided critical institutional support, as did the Fondation Alfred Kastler. At the University of Wisconsin-Madison, seed funding from the Center for Interdisciplinary French Studies, travel funds from the Graduate School Research Committee and the Global Studies research center, and a course buyout from the Division of International Studies allowed me to get the project going. Further resources from a WARF-H. I. Romnes faculty fellowship and from the Center for European Studies, as well as the ongoing support of the Department of Medical History and Bioethics, filled the gaps and allowed for a number of return travel opportunities to complete the fieldwork. Finally, the Andrew W. Mellon Sawyer Seminar on Biopolitics at the Center for the Humanities provided a critical intellectual space to think through this project's implications with a dazzling array of colleagues from Wisconsin and a number of other institutions.

In Paris, conversations with colleagues and public health professionals helped me to frame the project. At CERMES in Villejuif, I thank Isabelle Baszanger for her early support and for introducing me to Daniel Benamouzig, Martine Bungener, Jean-Paul Gaudillière, and Ilana Llowy, who in turn helped me meet Olivier Borraz and Danielle Salamon, all of whom helped me think about how qualitative social science was essential to the study of natural disaster. Robert Dingwall, Anne Murcott, and Joëlle Vailly provoked important conversations early in the project. Françoise Cribier and Elise Feller introduced me to the history and sociology of aging in France, which became central to the project. Anne Munck invited me to her home to discuss the experience of medical practice during the disaster; Patrick Pelloux and David Schnall also provided essential information on this subject. At the Atelier Parisien de Santé Publique, Alfred Spira and Emmanuelle Cadot offered keen insight into the medical geography of the disaster. Eric Jougla, Gérard Pavillon, Grégoire Rey, and Laurent Toulemon shared with me some of the finer points of the epidemiological and demographic study of disaster. Others who generously shared their time include Danièle Alet, Constance Fulda, Jean-François Mattei, François Michaud Nérard, Gregory Quenet, Cécile Rocca, and Brigitte Roux. Anne Véga kindly sent me a number of her publications on the anthropology of the heat wave. I am especially indebted to Didier Fassin, whose patronage allowed me to take a grant from the city of Paris at IRIS, and to Carine Vassy, who has been an exceptional collaborator and interlocutor.

I owe a number of deep intellectual debts to my colleagues at the University of Wisconsin-Madison. In October 2003, then-Dean Philip M. Farrell of the Medical School first planted the seed for this project when he passed me a note during a meeting suggesting I investigate the disaster—I owe him deeply for this. Then-Dean Gilles Bousquet of the Division of International Studies was another early champion of this project. In the Department of Medical History and Bioethics, Warwick Anderson, Tom Broman, Norm Fost, Dan Hausman, Linda Hogle, Judy Houck, Judy Leavitt, Sue Lederer, Gregg Mitman, Ron Numbers, Robert Streiffer, and Claire Wendland have offered many useful suggestions along the way. Thanks also to my colleagues in the Department of the History of Science, who have helped me to test the limits of what counts as history. In the Center for Culture, History, and Environment, Bill Cronon and Lynn Keller have patiently listened to early versions of various aspects of this project. Comments from colleagues in the Holtz Center for Science and Technology Studies, including Joan Fujimura, Daniel Kleinman, and Amit Prasad, helped me to think about how sociologists might receive this work. Jonathan Patz, Lori DiPrete Brown, and their colleagues at the Center for Sustainability and the Global Environment as well as the Global Health Institute have been extraordinarily supportive for years. Thanks to a Mellon Sawyer Seminar on Biopolitics, I was able to think through much of this project's foundation with Sara Guyer, whose insights have proven critical to the book. The help of project assistants in the Department of Medical History and Bioethics, including Bridget Collins, Vicki Daniel, Kristin Hamilton, Judith Kaplan, Emer Lucey, Sarah Potratz, Lynnette Regouby, Katie Robinson, Andrew Ruis, and Shannon Withycombe, has been extraordinary. Completing the book would have been unimaginable without the peerless administrative staff in our office, including Lori Brooks, Lorraine Rondon, Sharon Russ, JoAnn Steinich, and in particular Jean Von Allmen. The UW Cartography Lab helped with map production, and Jen Lucas provided essential last-minute troubleshooting help with the photographs.

Along the way I have had important opportunities to present various drafts of my work to audiences who have offered critical insight and useful criticism of this project. I thank the historians and social scientists who helped to shape this project in the course of talks at Cornell University, Northern Illinois University, Rutgers University, the University of Minnesota, the University of Wisconsin at Milwaukee, and Washington University in St. Louis, and in particular, Kalman Applbaum, Paul Brodwin, Jennifer Gunn, Beatrix Hoffman, Susan Jones, Michael Lynch, Sara Pritchard, and Corinna Treitel. I inflicted early versions of this material on colleagues at conferences including those of the American Anthropological Association, the American Association for the History of Medicine, the American Society for Environmental History, Dynamiques Urbaines et Enjeux Sanitaires, Eco-Health One, the Natural Hazards Workshop, the One Health Summit, the Society for French Historical Studies, and the Western Society for French History. I thank in particular Andrew Aisenberg, Jennifer Boittin, Caroline Ford, Kim Fortun, Jeffrey Jackson, Roxanne Panchasi, Dora Weiner, and Tamara Whited for their comments. In less formal settings, David Barnes, Ben Brower, Brady Brower, J. P. Daughton, Deborah Jenson, Rick Jobs, Suzanne Kaufman, the late Harry Marks, Randy Packard, Mary Louise Roberts, Naomi Rogers, Rebecca Scales, Joan Scott, Todd Shepard, Dan Sherman, Bonnie Smith, Scott Straus, and John Warner have offered important suggestions and support.

The review process has been critical to this project's success. I thank a number of readers of earlier drafts of all or part of this work. An earlier version of chapter 3 appeared as "Place Matters: Mortality, Space, and Urban Form in the 2003 Paris Heat Wave Disaster," in *French Historical Studies* 36, no. 2 (2013): 299–330, and appears here by permission of Duke University Press. I thank Elinor Accampo, Jeffrey Jackson, and two anonymous readers for their insights. At the University of Chicago Press, Karen Darling's support and sustained interest in the project have been essential. I thank her and Sophie Werely for ably shepherding me through the publication process, and

Marian Rogers for her expert copyediting of the manuscript. A special thanks goes to Eric Klinenberg and Dana Simmons for reading through the entire book more than once and offering incisive comments. I of course accept all responsibility for any of the book's remaining flaws.

Family and friends in Madison and beyond have been essential to the project's completion. My mother, Roberta Keller, has offered constant encouragement (not to mention unlimited babysitting). Brooke Keller has been an incomparable partner through the process, gladly accepting the disruptions to daily life that accompanied the long process of research and writing. It would have been impossible to complete this book without her love, her tireless support, and her encouragement. My son, Max, has proven an important partner in this project as well. He took his first steps in Paris while I was conducting this research, and accompanied me on a number of follow-up trips. He is without doubt the world's best traveling companion, and I dedicate this book to him.

Notes

Introduction

1. While this is the official policy, in practice, most bodies remain in the tombs for at least several years longer.

2. The heat wave is France's worst natural disaster when measured in terms of mortality; by contrast, the windstorms of 1999 remain the worst in terms of property damage and insured losses.

3. Acacio Pereira and Patrick Roger, "Canicule: En région parisienne, quelque 300 corps n'ont toujours pas été réclamés par leur famille," *Le Monde,* 26 August 2003.

4. Chapters 1 and 5 discuss death management and the disaster in much greater detail.

5. Jean-Marie Robine et al., "Death Toll Exceeded 70,000 in Europe during the Summer of 2003," *Comptes rendus: Biologies* 331 (2008): 171–78.

6. Alexandre Adveev et al., "Populations et tendances démographiques des pays européens (1980–2010)," *Population-F* 66, no. 1 (2011): 9–133; see 46.

7. *Eurostat Regional Yearbook 2013* (Luxembourg: Publications Office of the European Union, 2013), 49.

8. "La ola de calor de 2003 causó casi 13.000 muertes en España," 20minutos.es, 25 November 2005; "Italy Puts 2003 Heat Toll at 20,000," *International Herald Tribune,* 28 June 2005.

9. The keywords for these searches were the following: in Italian, *ondata di calore* and *ondata di caldo*; in Spanish, *ola de calor*; and in French, *canicule.*

10. Martine Perez, "Le mois d'août meurtrier avait fait 20.000 de morts en Italie; La vérité révélée avec deux ans de retard," *Le Figaro,* 28 June 2005, 8. It is notable that this story appeared in a right-leaning publication that had downplayed the disaster in France, and that the article misrepresents the Italian report. Perez asserts that ISTAT, the Italian statistics office, indicated 20,000 excess deaths during "the heat wave of August 2003"; the report in fact ascribes 20,000 excess deaths to heat during the entire summer of 2003.

11. Robine et al., "Death Toll," 174: according to the authors, France experienced 15,251 deaths in August, while Italy experienced 9,713, and Spain 6,461.

12. Jean-Pierre Dupuy, *Petite métaphysique des tsunamis* (Paris: Seuil, 2005).

13. Amartya Sen, *Poverty and Famines: An Essay on Entitlement and Deprivation* (Oxford: Oxford University Press, 1981); Mike Davis, *Late Victorian Holocausts: The El Niño Famines and the Making of the Third World* (London: Verso, 2002); Kim Fortun, *Advocacy after Bhopal: En-*

vironmentalism, Disaster, New Global Orders (Chicago: University of Chicago Press, 2001); Ted Steinberg, *Acts of God: The Unnatural History of Natural Disaster in America* (New York: Oxford University Press, 2006); Channa N. B. Bambaradeniya et al., *A Report on the Terrestrial Assessment of Tsunami Impacts on the Coastal Environment in Rekawa, Ussangoda and Kalametiya (RUK) Area of Southern Sri Lanka* (Colombo, Sri Lanka: IUCN-World Conservation Union Sri Lanka Country Office, 2005).

14. Eric Klinenberg, *Heat Wave: A Social Autopsy of Disaster in Chicago* (Chicago: University of Chicago Press, 2002).

15. Kai Erikson, *A New Species of Trouble: Explorations in Disaster, Trauma, and Community* (New York: Norton, 1994); Nancy Scheper-Hughes, *Death without Weeping: The Violence of Everyday Life in Brazil* (Berkeley: University of California Press, 1992).

16. Charles Rosenberg, "Cholera in Nineteenth-Century Europe: A Tool for Social and Economic Analysis," *Comparative Studies in Society and History* 8 (1966): 452–63; on 452.

17. See Committee on Science, Engineering, and Public Policy and the National Academies, "Disasters Roundtable on Disasters and Community Resilience" (report, 37th Annual Natural Hazards Research and Applications Workshop, Broomfield, Colorado, 16 July 2012). Also Louis Comfort et al., "Reframing Disaster Policy: The Global Evolution of Vulnerable Communities," *Environmental Hazards* 1 (1999): 39–44.

18. Lee Clarke, *Worst Cases: Terror and Catastrophe in the Popular Imagination* (Chicago: University of Chicago Press, 2006); S. L. Cutter, *American Hazardscapes: The Regionalization of Hazards and Disasters* (Washington, DC: Joseph Henry Press, 2001); Dennis Mileti, *Disasters by Design: A Reassessment of Natural Hazards in the United States* (Washington, DC: Joseph Henry Press, 1999); Richard A. Posner, *Catastrophe: Risk and Response* (New York: Oxford University Press, 2004); Ben Wisner et al., *At Risk: Natural Hazards, People's Vulnerability and Disasters* (New York: Routledge, 2004).

19. Ulrich Beck, *Risk Society: Towards a New Modernity* (London: Sage, 1992); Anthony Giddens, *The Consequences of Modernity* (Stanford, CA: Stanford University Press, 1990); Joost Van Loon, *Risk and Technological Culture: Towards a Sociology of Virulence* (New York: Routledge, 2002); Charles Perrow, *Normal Accidents: Living with High-Risk Technologies* (Princeton, NJ: Princeton University Press, 1999).

20. Perrow, *Normal Accidents*; Diane Vaughan, *The Challenger Launch Decision: Risky Technology, Culture, and Deviance at NASA* (Chicago: University of Chicago Press, 1997).

21. S. L. Cutter, "The Vulnerability of Science and the Science of Vulnerability," *Annals of the Association of American Geographers* 93, no. 1 (2003): 1–12; Susan L. Cutter, Bryan J. Boruff, and W. Lynn Shirley, "Social Vulnerability to Environmental Hazards," *Social Science Quarterly* 84, no. 2 (2003): 242–61; Mike Davis, *Ecology of Fear: Los Angeles and the Imagination of Disaster* (New York: Vintage, 1998); Ted Steinberg, "The Secret History of Natural Disaster," *Environmental Hazards* 3 (2001): 31–35.

22. These intersections have been at the core of recent work in environmental history, and particularly the environmental histories of France and of the modern city. On the importance of culture in the shaping of nature, see William Cronon, "Introduction: In Search of Nature," in *Uncommon Ground: Toward Reinventing Nature*, ed. William Cronon (New York: W. W. Norton, 1995), 23–56; Davis, *Ecology of Fear*; Davis, *Late Victorian Holocausts*; Michael Bess, *The Light-Green Society: Ecology and Technological Modernity in France, 1960–2000* (Chicago: University of Chicago Press, 2003); Caroline Ford, "Landscape and Environment in French Historical and Geographical Thought: New Directions," *French Historical Studies* 24 (2001): 125–34; and Ford, "Nature, Culture, and Conservation in France and Her Colonies, 1840–1940," *Past and Pres-*

ent 183 (2004): 173–98; Sara B. Pritchard, "Reconstructing the Rhône: The Cultural Politics of Nature and Nation in Contemporary France, 1945–1997," *French Historical Studies* 27 (2004): 765–99; E. C. Spary, *Utopia's Garden: French Natural History from the Old Regime to Revolution* (Chicago: University of Chicago Press, 2000); J. R. McNeill, "Observations on the Nature and Culture of Environmental History," *History and Theory* 42 (2003): 5–43; Martin V. Melosi, "Cities, Technological Systems, and the Environment," *Environmental History Review* 14, nos. 1–2 (1990): 45–64; Melosi, "The Place of the City in Environmental History," *Environmental History Review* 17, no. 1 (1993): 1–24; Christine Meisner Rosen and Joel Arthur Tarr, "The Importance of an Urban Perspective in Environmental History," *Journal of Urban History* 20, no. 3 (1994): 299–310; Jeffrey K. Stine and Joel A. Tarr, "At the Intersection of Histories: Technology and the Environment," *Technology and Culture* 39, no. 4 (1998): 601–40; Joel A. Tarr and Gabriel Dupuy, *Technology and the Rise of the Networked City in Europe and America* (Philadelphia: Temple University Press, 1988).

23. See in particular Foucault's *Discipline and Punish: A History of the Prison*, trans. Alan Sheridan (New York: Vintage, 1977), esp. 30–31.

24. Steinberg, "The Secret History of Natural Disaster."

25. T. H. Marshall, "Citizenship and Social Class," in *Inequality and Society*, ed. Jeff Manza and Michael Sauder (New York: Norton, 2009), 148–54; Bryan S. Turner, "Outline of a Theory of Citizenship," in *Citizenship: Critical Concepts*, ed. Bryan S. Turner and Peter Hamilton (New York: Routledge, 1994), 199–226.

26. Although it has involved those as well: see especially the debates between Rogers Brubaker, *Citizenship and Nationhood in France and Germany* (Cambridge, MA: Harvard University Press, 1992) and Yasemin Soysal, *The Limits of Citizenship: Migrants and Postnational Membership in Europe* (Chicago: University of Chicago Press, 1994).

27. Joan Wallach Scott, "French Universalism in the Nineties," *Differences* 15, no. 2 (2004): 32–53.

28. Robert Castel, *Les métamorphoses de la question sociale: Une chronique du salariat* (Paris: Fayard, 1995).

29. Michel Foucault, *The History of Sexuality*, vol. 1, *An Introduction*, trans. Robert Hurley (New York: Vintage, 1990), 138.

30. See Giorgio Agamben, *Homo Sacer: Sovereign Power and Bare Life*, trans. Daniel Heller-Roazen (Stanford, CA: Stanford University Press, 1998).

31. Nikolas Rose, *The Politics of Life Itself: Biomedicine, Power, and Subjectivity in the Twenty-First Century* (Princeton, NJ: Princeton University Press, 2007), 53–54.

32. Mika Ojakangas, "Impossible Dialogue on Bio-Power: Agamben and Foucault," *Foucault Studies* 2 (2005): 5–28.

33. Peter Redfield, *Life in Crisis: The Ethical Journey of Doctors without Borders* (Berkeley: University of California Press, 2012); João Biehl, "Vita: Life in a Zone of Social Abandonment," *Social Text* 19, no. 3 (2001): 131–49.

34. Judith Butler, *Frames of War: When Is Life Grievable?* (London: Verso, 2009), 77.

35. Judith Butler, *Precarious Life: The Powers of Mourning and Violence* (London: Verso, 2004), xiv, 37–38.

36. Butler, *Frames of War*, 75.

37. Agamben, *Homo Sacer*, 127–28.

38. François Ewald, *L'État-providence* (Paris: Grasset, 1986). On the relationship between biopolitics and the neoliberal market, see also Melinda Cooper, *Life as Surplus: Biotechnology and Capitalism in the Neoliberal Era* (Seattle: University of Washington Press, 2008); also

Kaushik Sunder Rajan, *Biocapital: The Constitution of Postgenomic Life* (Durham, NC: Duke University Press, 2006); Stephen J. Collier, "Topologies of Power: Foucault's Analysis of Political Government beyond Governmentality," *Theory, Culture, and Society* 26 (2009): 78–108; and Nikolas Rose, Pat O'Malley, and Mariana Valverde, "Governmentality," *Annual Review of Law and Social Science* 2 (2006): 83–104. On the links between the notion of the family and the social in France, see Camille Robcis, *The Law of Kinship: Anthropology, Psychoanalysis, and the Family in France* (Ithaca, NY: Cornell University Press, 2013).

39. Robert Proctor, "Agnotology: A Missing Term to Describe the Cultural Production of Ignorance (and Its Study), in *Agnotology: The Making and Unmaking of Ignorance*, ed. Robert N. Proctor and Londa Schiebinger (Stanford, CA: Stanford University Press, 2008), 1–33.

40. Londa Schiebinger, "West Indian Abortifacients and the Making of Ignorance," in Proctor and Schiebinger, *Agnotology*, 149–62.

41. Ibid., 152.

42. Jeanne Guillemin, *Anthrax: The Investigation of a Deadly Outbreak* (Berkeley: University of California Press, 1999).

Chapter One

1. Julian Jackson, *The Popular Front in France: Defending Democracy, 1934–38* (New York: Cambridge University Press, 1988), 132.

2. Ellen Furlough, "Making Mass Vacations: Tourism and Consumer Culture in France, 1930s to 1970s," *Comparative Studies in Society and History* 40, no. 2 (1998): 247–86. This was of course not exclusively a French phenomenon: see also Orvar Löfgren, *On Holiday: A History of Vacationing* (Berkeley: University of California Press, 1999); and Shelley Baranowski, *Strength through Joy: Consumerism and Mass Tourism in the Third Reich* (New York: Cambridge University Press, 2007).

3. TF1, *20 heures*, 1 August 2003.

4. Nathalie Rihouet, France 2, *20 heures—Le journal*, 1 August 2003.

5. Catherine Laborde, TF1, *Météo*, 1 August 2003.

6. Laurent Chabrun et al., "Qui étaient les oubliés de la canicule?," *L'Express*, 25 December 2003.

7. See, e.g., Bob Henson, "The Weather Notebook," interview with Eric Klinenberg, 8 July 2004, http://www.weathernotebook.org/transcripts/2004/07/08.php.

8. TF1, *20 heures*, 2 August 2003.

9. TF1, *20 heures*, 3 August 2003 and 9 August 2003.

10. TF1, *20 heures*, 6 August 2003; France 2, *20 heures—Le journal*, 11 August 2003.

11. France 2, *20 heures—Le journal*, 9 August 2003.

12. TF1, *20 heures*, 8 August 2003.

13. TF1, *20 heures*, 7 August 2003.

14. TF1, *20 heures*, 8 August 2003.

15. TF1, *20 heures*, 2 August 2003.

16. Météo-France, "Conseils de comportement mis sur le site meteo.fr pendant la canicule," in Assemblée Nationale, No. 1091, *Rapport d'information sur la crise sanitaire et sociale déclenchée par la canicule*, 2 vols., 24 September 2003 (henceforth "Rapport Jacquat"), Annexe 21: Documents remis à l'appui de l'audition de Michèle Froment-Vedrine et Jean-Pierre Beysson.

17. E-mail from Loic Josseran to Renée Pomarède, 7 August 2003; in Rapport Jacquat, Annexe 5/1: Documents remis à l'appui de l'audition de Gilles Brucker.

18. E-mail from William Dab to Yves Coquin, 6 August 2003; in Rapport Jacquat, Annexe 11/1: Documents remis à l'appui de l'audition de William Dab.

19. E-mail from Marc Verny to Ghislaine Calavia and Yves Coquin, 8 August 2003; in Rapport Jacquat, Annexe 11/1.

20. E-mail from Jean-Claude Desenclos to Hubert Isnard and Yves Coquin, 8 August 2003; in Rapport Jacquat, Annexe 11/1.

21. Direction Générale de la Santé, "Fortes chaleurs en France : Recommandations sanitaires," 8 August 2003; in Rapport Jacquat, Annexe 11/1.

22. Charles de Saint-Sauveur, "Canicule: Les victimes de la chaleur de plus en plus nombreuses," *Le Parisien*, 10 August 2003.

23. "RAS ('rien à signaler') dans la main courante": see "Institut de veille sanitaire et crise liée à la canicule: Chronologie," in Rapport Jacquat, Annexe 5/1: Documents remis à l'appui de l'audition de Gilles Brucker.

24. See Delphine Brard, "La fabrique médiatique de la canicule d'août 2003 comme problème public" (DEA mémoire, Université de Paris-I Panthéon-Sorbonne, 2004).

25. TF1, *20 heures*, 2 August 2003.

26. See, e.g., TF1, *20 heures*, 2 August 2003; and France 2, *20 heures—Le journal*, 8 August 2003.

27. TF1, *Météo*, 3 August 2003; and *20 heures*, 3 August 2003 and 8 August 2003.

28. TF1, *20 heures*, 8 August 2003.

29. E-mail from Yves Coquin to Jean-Claude Desenclos and Renée Pomarède, 8 August 2003; in Rapport Jacquat, Annexe 11/1.

30. Marc Payet, in Rapport Jacquat, vol. 2, p. 444; also Marc Payet and Charles de Saint-Sauveur, "Canicule: L'histoire," *Le Parisien*, 14 August 2003.

31. Jacques Kerdoncuff, in Assemblée Nationale, No. 1455, *Rapport fait au nom de la commission d'enquête sur les conséquences sanitaires et sociale de la canicule*, 2 vols., 25 February 2004 (henceforth "Rapport Evin"), vol. 2, p. 45.

32. TF1, *20 heures*, 10 August 2003.

33. The Association des Médecins Urgentistes Hospitaliers de France.

34. For example, in May and June of 2003, he accused regional health authorities of setting the stage for disaster with promised budget cuts for the summer, and in late July, he had insisted to the director of Paris's hospital system that these cuts had already brought emergency and critical care facilities to a state of "engorgement" and that the situation was "unmanageable." See Patrick Pelloux, in Rapport Evin, vol. 2, pp. 218–19.

35. Cyrille Louis, a health journalist for the Parisian daily *Le Figaro*, cited in Brard, "La fabrique médiatique de la canicule," 39. For a more measured critique of Pelloux, see the testimony of Rose-Marie van Lerberghe, director of Paris's hospital authority, who noted that she was contacted by journalists about so-called overcrowding scandals in Paris wards, only to discover large numbers of available beds; Rapport Evin, vol. 2, p. 111.

36. TF1, *20 heures*, 11 August 2003.

37. See, e.g., Paul Rabinow, *French DNA* (Chicago: University of Chicago Press, 1997), on related issues concerning the politics of life in French biotechnology in this period. These were of course highly controversial issues in their own right, as detailed by Robcis, *The Law of Kinship*.

38. Jean-Louis Sanmarco, cited in Patrick Pelloux, *Urgentiste* (Paris: Fayard, 2004), 16.

39. TF1, 20 heures, 11 August 2003.

40. See Brard, "La fabrique médiatique de la canicule," 49–62.

41. "Le petit polo de Mattei," *Le Parisien*, 31 December 2003. For his part, Mattei argues that there was no real divide between the hospitals' and the administration's experience of the crisis, asserting that Pelloux was the only practitioner complaining about conditions and that even he (Pelloux) was underestimating the crisis with his comments about "fifty deaths"; interview with Mattei, 11 October 2011.

42. Marc Payet, in Rapport Jacquat, vol. 2, p. 445.

43. http://societe.fluctuat.net/jean-francois-mattei.html.

44. Eric Favereau, "Mattei, ministre en mauvais santé," *Libération*, 21 August 2003.

45. DGS, "Saturation des chambres funéraires ou mortuaires en Île de France," communiqué de presse, in Rapport Jacquat, Annexe 11/3.

46. Jean Carlet to Lucien Abenhaïm, 11 August 2003, in Rapport Jacquat, Annexe 11/2. I have translated *réa*, short for *réanimation*, as "ICU."

47. Joëlle Le-Moal to Yves Coquin, 12 August 2003, in Rapport Jacquat, Annexe 11/2.

48. Fiche Technique, Institut Médico-Légal de Paris; fax from Yves Coquin to Eric Freysselinard, 12 August 2003; and fax from Eric Freysselinard to Yves Coquin, 12 August 2003, all in Rapport Jacquat, Annexe 11/3.

49. France 2, *20 heures—Le journal*, 12 August 2003.

50. TF1, *20 heures*, 13 August 2003.

51. Rapport Jacquat, Annexe 5/1: Chronique InVS.

52. Danielle Golinelli to Yves Coquin, 13 August 2003; memorandum from Direction de l'Hospitalisation et de l'Organisation des Soins to Mme Toupiller, in Rapport Jacquat, Annexe 11/4.

53. Pelloux, *Urgentiste*, 75–76.

54. France 2, *20 heures—Le journal*, 13 August 2003.

55. TF1, *20 heures*, 13 August 2003.

56. Rapport Jacquat, vol. 2, p. 107.

57. Vérane Castelnau et al., "De l'hôpital au cimetière, la saturation," *Libération*, 14 August 2003.

58. Cited in Catherine Le Grand-Sébille and Anne Véga, *Pour une autre mémoire de la canicule: Professionnels de funéraire, des chambres mortuaires et familles témoignent* (Paris: Vuibert, 2005), 63.

59. Isabelle Thirouin to Yves Coquin, 14 August 2003, in Rapport Jacquat, Annexe 11/6.

60. Isabelle Thirouin to Yves Coquin, 14 August 2003, in Rapport Jacquat, Annexe 11/8.

61. Direction Générale de la Santé, Note pour le ministre à l'attention de Mme A. Bolot-Gittler, Directrice adjointe du cabinet, 16 August 2003.

62. Cited in Le Grand-Sébille and Véga, *Pour une autre mémoire de la canicule*, 40–41.

63. Le Grand-Sébille and Véga, *Pour une autre mémoire de la canicule*, 42–44.

64. Pascale Robert-Diard, "Le cadavre, la cour d'appel et le trouble anormal de voisinage," *Le Monde*, 18 May 2009.

65. Le Grand-Sébille and Véga, *Pour une autre mémoire de la canicule*, 54.

66. See, e.g., TF1, *20 heures*, 14 August 2003.

67. Spike Jean, Alexandre Aïchouba, Loïc de la Mornais, Eric Delagneau, and Cécile Henry, France 2, *La mort en face*, 4 December 2003.

68. "Fortes chaleurs en France: Aspects sanitaires et recommandations," attachment to e-mail from CAB-SANTE-PRESSE to Yves Coquin and Danielle Toupillier, 11 August 2003, in Rapport Jacquat, Annexe 11/1. See also Pelloux, *Urgentiste*, 72.

69. France 2, *20 heures—Le journal*, 11 August 2003.

70. France 2, *20 heures—Le journal*, 13 August 2003.

71. France 2, *20 heures—Le journal*, 15 August 2003.

72. "Sanitaire sans précédent," *Le Parisien*, 15 August 2003, 7.

73. France 2, *20 heures—Le journal*, 17 August 2003.

74. Cited in Dominique Delpiroux, "La canicule tue, la polémique s'installe," *La Dépêche*, 12 August 2003.

75. France 2, *20 heures—Le journal*, 15 August 2003.

76. Olivier Aubry, "Les 35 heures ont aggravé les tensions dans les hôpitaux," *Le Parisien*, 14 August 2003.

77. Pelloux, *Urgentiste*, 60–61.

78. Vanessa Schneider, "Le gouvernement poursuit la chasse aux fainéants," *Libération*, 29 August 2003.

79. France 2, live coverage of president's speech, 21 August 2003.

80. TF1, *20 heures*, 21 August 2003.

81. France 2, *20 heures—Le journal*, 21 August 2003.

82. Pierre Marcelle, "D'un été meurtrier," *Libération*, 25 August 2003.

83. Jean-Marc Roubaud, cited in Rapport Evin, vol. 2, p. 430.

84. See, e.g., Lucien Abenhaïm, *Canicules: La santé publique en question* (Paris: Fayard, 2003), but also France 2, *20 heures—Le journal*, 18 August 2003.

85. Thierry Boudes and Hervé Laroche, "Taking Off the Heat: Narrative Sensemaking in Post Crisis Inquiry Reports," *Organization Studies* 30 (2009): 377–96.

86. See, e.g., La chaine parlementaire, *Mission d'information sur les conséquences de la canicule*, 11 September 2003, or much of the testimony in Rapport Evin, vol. 2.

87. Françoise Lalande et al., *Mission d'expertise et d'évaluation du système de santé pendant la canicule de 2003*, September 2003, 5 (henceforth "Lalande Report").

88. Lalande Report, 2.

89. Ibid., 40.

90. Critics raised these and other issues immediately upon the report's release; see Sandrine Cabut, Eric Favereau, and Julie Lasterade, "Les extraits du rapport et les réponses des professionnels concernés: 'Il ne faut pas chercher des boucs émissaires'," *Libération*, 9 September 2003. See also La chaine parlementaire, *Mission d'information sur les conséquences de la canicule*, 11 September 2003, and in particular the interrogation of Mattei by Communist deputy Maxime Gremetz, who questioned the validity of the report.

91. Denis Hémon and Eric Jougla, *Surmortalité liée à la canicule d'août 2003—Rapport d'étape: Estimation de la surmortalité et principales caractéristiques épidémiologiques* (Paris: Inserm, 2003), 54 (henceforth *Surmortalité—Rapport*).

92. The report is a careful and deliberate study of available data, and several interviews I conducted with Jougla in 2007 reinforced my sense of the report's professionalism.

93. See chapter 5 for an extended evaluation of this problem.

94. Perrow, *Normal Accidents*, 5–8.

Chapter Two

1. Clara Dupont-Monod, "Le secret de Marie France," *Marianne* 334 (15–21 September 2003): 22–23.

2. For reasons of confidentiality, I have concealed the identity of most informants in the citations, as well as the complete street address where the death took place. In most cases I have

also noted only the given name for victims of the heat wave; exceptions are those whom the French media have already profiled with their full names.

3. "Canicule: La liste des 66 morts oubliés," *Le Parisien*, 2 September 2003.

4. "450 morts oubliés: Tous coupables," *Le Parisien*, 26 August 2003; "Nous devrions avoir honte," *Le Parisien*, 26 August 2003; "À ses morts oubliés, la patrie repentante," *Libération*, 4 September 2003; "Barbarie française," *Le Figaro*, 25 August 2003; "Quand l'égoïsme et l'indifférence ont tué," *Le Figaro*, 28 August 2003.

5. François Michaud Nérard, *La révolution dans la mort* (Paris: Vuibert, 2007), 79–80.

6. Current law stipulates that funeral services are provided for free "to persons deprived of sufficient resources"; see article L2223–27 of the Code général des collectivités territoriales.

7. http://www.landrucimetieres.fr/spip/spip.php?article1858.

8. The cemetery also receives monthly, on average, a dozen stillborn infants and nonviable newborns who die in their first few days in the hospital, but it buries them in traditional earth-bound graves.

9. At one point in late August the cemetery and its temporary storage centers held some 300 bodies; the city's genealogical services managed to place nearly two-thirds of these bodies with family or friends before the 3 September ceremony. Staff buried many in late August. Nearly 60 remained unburied on the day of the ceremony.

10. Klinenberg, *Heat Wave*, 236–40.

11. "Le cimetière en état de siège," *Le Parisien*, 4 September 2003.

12. Forty others had been buried in the weeks before 3 September, and therefore figured neither on the list nor in the official ceremony. Interview with funeral official, Paris, 28 June 2011.

13. TF1, *20 heures*, 3 September 2003.

14. France 2, *20 heures—Le journal*, 3 September 2003.

15. Didier Arnaud, "À ses morts oubliés, la patrie repentante," *Libération*, 4 September 2003.

16. Marc Payet, "Le dernier hommage aux oubliés de la canicule," *Le Parisien*, 4 September 2003, 10.

17. Marc Payet, "Je tenais à être là," *Le Parisien*, 4 September 2003, 10.

18. Dominique Quinio, "Aux morts inconnus," *La Croix*, 3 September 2003.

19. "Courrier," *La Croix*, 13 October 2003.

20. Solenn de Royer, "Les 'oubliés de la canicule' avaient presque tous de la famille," *La Croix*, 20 October 2003.

21. Michel Waintrop, "Au 'carré des indigents,' des hommes et des numéros," *La Croix*, 3 November 2003.

22. France 3, *Vieillir ensemble?*, 23 September 2003.

23. *Les oubliés de la canicule*, directed by Sophie Lepault and Ibar Aibar (Paris: Doc en Stock, 2003–4).

24. Ibid.

25. Interview with Constance Fulda, sixteenth arrondissement, Paris, 18 June 2009. I also cite her notes on the exhibition.

26. Fulda was kind enough to give me two of the pieces.

27. See http://constance.fulda.pagesperso-orange.fr/#. The screen was also exhibited in Barbizon in March 2004, at the Galerie Joyce in Paris in May 2006, and at the Centre Culturel in Melun in July 2007.

28. Quinio, "Aux morts inconnus."

29. Simon Marty, "Philippe Heurteaux, le clochard millionaire," *Marianne* 334 (15–21 September 2003): 26–27.

30. Anna Alter and Perrine Cherchève, "Mihajlo Molerovic, radié de la vie," *Marianne* 334 (15–21 September 2003): 24–25.

31. Dupont-Monod, "Le secret de Marie France."

32. Raphaël Chamak, Alexandre Garcia, and Cécile Prieur, "Ces vieillards morts dans l'oubli à Paris lors de la canicule," *Le Monde*, 4 September 2003.

33. Laurent Chabrun et al., "Qui étaient les oubliés de la canicule?"

34. Some accounts say fifty-eight, but fifty-seven is the consensus.

35. For details about where they lived, see chapter 3.

36. Fifteen of the victims had unknown marital status at the time of their deaths.

37. It is impossible to discern their ethnicity, as such data are not included in civil records (and are indeed illegal to collect in France).

38. Klinenberg, *Heat Wave*, 74–77.

39. Clifford Geertz, *The Interpretation of Cultures* (New York: Basic Books, 1973), 3–30 and 412–53.

40. Ibid., 9 and 20.

41. See Catherine Gallagher and Stephen Greenblatt, *Practicing New Historicism* (Chicago: University of Chicago Press, 2001), 28.

42. Joel Fineman, "The History of the Anecdote: Fiction and Fiction," in *The New Historicism*, ed. Harold Aram Veeser (New York: Routledge, 1989), 49–76.

43. Stephen Greenblatt, "The Touch of the Real," *Representations* 59 (1997): 14–29.

44. Gallagher and Greenblatt, *Practicing New Historicism.*

45. And not in 1979 as Gallagher and Greenblatt have it; see *Practicing New Historicism,* 66–74.

46. Michel Foucault, "La vie des hommes infâmes," in *Dits et écrits, 1954–1988*, vol. 3, ed. Daniel Defert and François Ewald (Paris: Gallimard, 1994), 237–53.

47. Ibid., 238.

48. Ibid., 239.

49. See also similar details in Dupont-Monod, "Le secret de Marie France."

50. *Aux oubliés de la canicule*, directed by Danièle Alet (France, 2004).

51. Pierre Bourdieu, "L'illusion biographique," *Actes de la recherche en sciences sociales* 62–63 (1986): 69–72.

Chapter Three

1. See Stephanie Vandentorren et al., "Données météorologiques et enquêtes sur la mortalité dans 13 grandes villes françaises," *Bulletin épidémiologique hebdomadaire* 45–46 (2003): 219–20; and Vandentorren et al., "Mortality in 13 French Cities during the August 2003 Heat Wave," *American Journal of Public Health* 94, no. 9 (2004): 1518–20. "Excess mortality" is a blunt but commonly accepted measurement for disaster-related mortality. It is calculated by subtracting the average mortality for a given period (in this case, 1–20 August, 2000–2002) from the observed mortality in the event of the disaster (1–20 August 2003). For more on the measurement of mortality during the disaster, see Carine Vassy, Richard Keller, and Robert Dingwall, *Enregistrer les morts, identifier les surmortalités: Une comparaison Angleterre, États-Unis et France* (Rennes: Presses de l'EHESP, 2010).

2. Pierre Bourdieu, "Effets de lieu," in *Le misère du monde*, ed. Pierre Bourdieu (Paris: Seuil, 1993), 159–67; citation on 159–60.

3. For a few examples of the qualitative study of disaster and vulnerability, see Jeffrey H.

Jackson, *Paris under Water: How the City of Light Survived the Great Flood of 1910* (London: Palgrave Macmillan, 2010); Davis, *Late Victorian Holocausts;* Davis, *Ecology of Fear*; Steinberg, *Acts of God*; Steinberg, "The Secret History of Natural Disaster"; and Sen, *Poverty and Famines.*

4. See Éloi Laurent, "Bleu, Blanc . . . Green? France and Climate Change," *French Politics, Culture, and Society* 27 (2009): 142–53; Le Grand-Sébille and Véga, *Pour une autre mémoire de la canicule*; and Martine Bungener, "Canicule estivale: La triple vulnérabilité des personnes âgées," *Mouvements* 32 (2004): 75–82.

5. The other two were homeless victims who died in hospitals, with no indication of their ordinary sites of residence, whereas several other homeless victims died in the street, and city officials noted the location of death on the death notice.

6. I borrow the term from Judith Butler's series of essays *Precarious Life* (London: Verso Press, 2004), in which she discusses the ways in which certain deaths, assimilable to our own experiences, are "grievable," while others are not so. Rhetorical processes of dehumanization accompany extraordinary physical and epistemological violence, thereby placing certain populations at enormous risk while sparing others. Butler's argument concerns the global war on terror, but the metaphor of precarious life and the processes of dehumanization it encompasses aptly suits the cases of these disaster victims.

7. On the daily distribution of deaths during the heat wave, see Hémon and Jougla, *Surmortalité—Rapport*, 29.

8. See http://www.insee.fr/fr/bases-de-donnees/default.asp?page=statistiques-locales/chiffres-cles.htm.

9. There are no explicit studies about heat and homelessness for 2003. However, only four of the ninety-five forgotten were homeless. In a conversation in 2009, Cécile Rocca, the director of Morts de la Rue, an NGO that organizes funerals and burials for France's homeless, told me that contrary to popular belief, relatively few homeless deaths are attributable to weather extremes. Instead, violence, infectious disease, and complications from mental illness and addiction are far more common causes of death.

10. E.g., Hémon and Jougla, *Surmortalité—Rapport*; also Alain Le Tertre et al., "Impact of the 2003 Heat Wave on All-Cause Mortality in 9 French Cities," *Epidemiology* 17, no. 1 (2006): 75–79; and Lalande Report.

11. See, for example, Institut de Veille Sanitaire, *Étude des facteurs de risque de décès des personnes âgées résidant à domicile durant la vague de chaleur d'août 2003* (Paris: InVS, 2004). Also Joël Belmin, "Les conséquences de la vague de chaleur d'août 2003 sur la mortalité des personnes âgées: Un premier bilan," *La presse médicale* 32 (18 Oct. 2003): 1591–94.

12. Of the official tally of 14,802 excess deaths ascribed to the 2003 heat wave, the demographers Denis Hémon and Eric Jougla list 4866.9, or 32.9 percent, in the Île-de-France; see Hémon and Jougla, *Surmortalité—Rapport*, 32.

13. These figures are based on data from the 1999 census collected by the Institut National de la Statistique et des Études Économiques (Insee), and consider the relationship of the city of Paris proper (population 2,125,851) to metropolitan France (population 58,520,688). The epidemiologists Florence Canouï-Poitrine, Emmanuelle Cadot, and Alfred Spira of the Atelier Parisien de Santé Publique calculated mortality in Paris at 1,067 excess deaths, or nearly 7.2 percent of the total of 14,802. See Canouï-Poitrine, Cadot, and Spira, "Excess Deaths during the August 2003 Heat Wave in Paris, France," *Revue d'épidémiologie et de santé publique* 54 (2006): 127–35.

14. For a definition of the urban heat island effect, see http://www.epa.gov/heatisld/index.html. For links between the heat island effect and heat waves in France in particular, see Clé-

ment Champiat, "Identifier les îlots de chaleur urbains pour réduire l'impact sanitaire des vagues de chaleur," *Environnement, risques et santé* 8, no. 5 (2009): 399–411.

15. For the eighteenth century, for example, see Louis-Sébastien Mercier, *Tableau de Paris* (Hamburg: Virchaux; Neuchâtel: S. Fauche, 1781); for urban pathology in the nineteenth and early twentieth centuries, see Gregg Mitman, "In Search of Health: Landscape and Disease in American Environmental History," *Environmental History* 10 (2005): 184–210; Charles Rosenberg, "Pathologies of Progress: The Idea of Civilization as Risk," *Bulletin of the History of Medicine* 72 (1998): 714–30; Judith Walkowitz, *City of Dreadful Delight: Narratives of Sexual Danger in Late Victorian London* (Chicago: University of Chicago Press, 1992). For examples of urban social pathology in mid-twentieth-century France and the United States, see Haut Comité Consultatif de la Population et de la Famille, *Politique de la vieillesse: Rapport de la commission d'étude des problèmes de la vieillesse* (Paris: La Documentation française, 1962), which attributes the increasing isolation of the elderly in France to the egoism and alienation of urban life; and Stanley Milgram, "The Experience of Living in Cities," *Science* 167, no. 3924 (1970): 1461–68.

16. Karen E. Smoyer, "Putting Risk in Its Place: Methodological Considerations for Investigating Extreme Event Health Risk," *Social Science and Medicine* 47, no. 11 (1998): 1809–24.

17. Klinenberg, *Heat Wave.*

18. Canouï-Poitrine, Cadot, and Spira, "Excess Deaths"; see also Emmanuelle Cadot and Alfred Spira, "Canicule et surmortalité à Paris en août 2003, le poid des facteurs socio-économiques," *Espaces, populations, sociétés* 2–3 (2006): 239–49.

19. As some of these deaths occurred in hospitals, the APSP's studies focused on those who died at home: including hospital deaths would exaggerate the death rate in arrondissements with hospitals. Moreover, the rate of those dying at home was far higher than that of those dying in hospitals: while in any given August, some 200 Parisians die in their homes, in 2003 roughly 900 died at home. Conversations with Alfred Spira and Emmanuelle Cadot, June 2005 and March 2007.

20. Each of Paris's twenty arrondissements is divided into four administrative *quartiers.* Cadot and Spira, "Canicule et surmortalité à Paris en août 2003."

21. Louis-René Villermé, "De la mortalité dans les divers quartiers de la ville de Paris," *Annales d'hygiène publique et de médecine légale* 3 (1830): 294–341; cf. Joshua Cole, *The Power of Large Numbers: Population, Politics, and Gender in Nineteenth-Century France* (Ithaca, NY: Cornell University Press, 2000); David Barnes, *The Making of a Social Disease: Tuberculosis in Nineteenth-Century France* (Berkeley: University of California Press, 1995); Barnes, *The Great Stink of Paris and the Nineteenth-Century Struggle against Filth and Germs* (Baltimore: Johns Hopkins University Press, 2006); Alain Corbin, *Le miasme et la jonquille* (Paris: Flammarion, 1982); William Coleman, *Death Is a Social Disease: Yellow Fever in the North* (Madison: University of Wisconsin Press, 1982).

22. Cole, *The Power of Large Numbers.*

23. Coleman, *Death Is a Social Disease.*

24. See Barnes, *The Great Stink of Paris,* 50; and Anthony Sutcliffe, *The Autumn of Central Paris: The Defeat of Town Planning, 1850–1970* (London: Edward Arnold, 1970), 102–3. Also David P. Jordan, *Transforming Paris: The Life and Labors of Baron Haussmann* (New York: Free Press, 1995), esp. 48, 96–97, 270–74; also David Pinkney, *Napoleon III and the Rebuilding of Paris* (Princeton, NJ: Princeton University Press, 1958), 30–31.

25. Jordan, *Transforming Paris*; Jordan, "Haussmann and Haussmannization: The Legacy for Paris," *French Historical Studies* 27, no. 1 (2004): 87–113; Pinkney, *Napoleon III*; Barnes, *The Great Stink of Paris*; T. J. Clark, *The Painting of Modern Life: Paris in the Art of Manet and His*

Followers (Princeton, NJ: Princeton University Press, 1985); Jeanne Gaillard, *Paris, la ville (1852–1870)* (Paris: L'Harmattan, 1977); David Harvey, *Consciousness and the Urban Experience: Studies in the History and Theory of Capitalist Urbanization* (Baltimore: Johns Hopkins University Press, 1985); André Morizet, *Du vieux Paris au Paris moderne* (Paris: Hachette, 1932).

26. Matthew Gandy, "The Paris Sewers and the Rationalization of Urban Space," *Transactions of the Institute of British Geographers*, n.s. 24 (1999): 23–44.

27. Pinkney, *Napoleon III*, 127.

28. Clark, *The Painting of Modern Life.*

29. Marshall Berman, *All That Is Solid Melts into Air: The Experience of Modernity* (New York: Penguin, 1982).

30. See Tyler Stovall, *The Rise of the Paris Red Belt* (Berkeley: University of California Press, 1990); Stovall, "From Red Belt to Black Belt: Race, Class, and Urban Marginality in Twentieth-Century Paris," in *The Color of Liberty: Histories of Race in France*, ed. Sue Peabody and Tyler Stovall (Durham, NC: Duke University Press, 2003), 351–70; Barnes, *The Great Stink of Paris*; Berman, *All That Is Solid Melts into Air*; Clark, *The Painting of Modern Life*; and Susan Buck-Morss, *The Dialectics of Seeing: Walter Benjamin and the Arcades Project* (Cambridge, MA: MIT Press, 1991).

31. For an account of what life might have been like in such a building before Haussmann's time, melding labor and domestic life, see Arlette Farge, *Fragile Lives: Violence, Power, and Solidarity in Eighteenth-Century Paris*, trans. Carole Shelton (Cambridge, MA: Harvard University Press, 1993), esp. 9–21.

32. E.g., Pinkney, *Napoleon III*, 8–9.

33. See Clark, *The Painting of Modern Life*; Ann-Louise Shapiro, *Housing the Poor of Paris: 1850–1902* (Madison: University of Wisconsin Press, 1985); Sutcliffe, *The Autumn of Central Paris.*

34. Sutcliffe, *The Autumn of Central Paris*; Shapiro, *Housing the Poor of Paris.*

35. Pinkney, *Napoleon III*, 91–92.

36. Shapiro, *Housing the Poor of Paris*, 36.

37. Ibid., 104.

38. See, e.g., Paul Auster, *The Invention of Solitude* (New York: Penguin, 1982); Philippe Bourgois, "Missing the Holocaust: My Father's Account of Auschwitz from August 1943 to June 1944," *Anthropological Quarterly* 78, no. 1 (2005): 89–123; Stephen Fredman, "'How to Get Out of the Room That Is the Book?' Paul Auster and the Consequences of Confinement," *Postmodern Culture* 6, no. 3 (1996), muse.jhu.edu/journals/postmodern_culture/v006/6.3fredman.html; Christine Coffman, "The Papin Enigma," *GLQ: A Journal of Lesbian and Gay Studies* 5, no. 3 (1999): 331–59; Kristin Ross, "Schoolteachers, Maids, and Other Paranoid Histories," *Yale French Studies* 91 (1997): 7–27. For a contemporary glimpse into living conditions in such rooms, see the documentary film *Chambre de bonne*, directed by Maija-Lene Rettig (Germany, 2002); far more problematic is the recent comedy *Les femmes du 6e étage*, directed by Philippe Le Guay (France, 2011), which trivializes power relations between a bourgeois in the sixteenth arrondissement and his domestic servant who lives in a *chambre de bonne.*

39. Few scholars have engaged the concept of vertical geographies. Most of those who have examined the problem have focused on high-rises. See, e.g., Donald McNeill, "Skyscraper Geography," *Progress in Human Geography* 29, no. 1 (2005): 41–55; and Jane M. Jacobs, Stephen Cairns, and Ignaz Strebel, "'A Tall Storey . . . but, a Fact Just the Same': The Red Road High-Rise as a Black Box," *Urban Studies* 44, no. 3 (2007): 609–29. The geographer Matthew Gandy has also hinted at the importance of verticality as a dimension in the contemporary city; see Gandy,

"Zones of Indistinction: Bio-Political Contestations in the Urban Arena," *Cultural Geographies* 13 (2006): 497–516; Gandy, "Cyborg Urbanization: Complexity and Monstrosity in the Contemporary City," *International Journal of Urban and Regional Research* 29, no. 1 (2005): 26–49; and Gandy, "The Ecological Facades of Patrick Blanc," *Architectural Design* 80, no. 3 (2010): 28–33. One particularly interesting example is Eyal Weizman, an Israeli architect, whose article series "The Politics of Verticality" appeared on the website OpenDemocracy.org (http://www.opendemocracy.net/conflict-politicsverticality/article_801.jsp). Weizman argues that there is a three-dimensionality to the uses of space in the Israeli-Palestinian conflict, and points to a biopolitical architecture of warfare that structures vulnerability against Palestinians through the vertical organization of space. Thanks to Kris Olds for pointing me to a number of these sources.

40. See especially Foucault, *The History of Sexuality*, vol. 1; also, with regard to sovereignty and territoriality, see Foucault, "Governmentality," trans. Rosi Braidotti and rev. Colin Gordon, in *The Foucault Effect: Studies in Governmentality*, ed. Graham Burchell, Colin Gordon, and Peter Miller (Chicago: University of Chicago Press), 87–104.

41. As per 2008 Insee data: see http://www.insee.fr/fr/bases-de-donnees/esl/resume.asp?codgeo=75115&nivgeo=arm.

42. The sixteenth had in 2008 a net household income of 77,085 euros, or more than triple the eighteenth's of 24,229 euros in the same year. See http://www.insee.fr/fr/bases-de-donnees/default.asp?page=statistiques-locales/chiffres-cles.htm.

43. Institut de Veille Sanitaire, *Étude des facteurs de risque.*

44. Ibid., 60.

45. Code de la construction et de l'habitation, Art. 111–6-1.

46. Décret n°2002–120 du 30 janvier 2002 relatif aux caractéristiques du logement décent pris pour l'application de l'article 187 de la loi n° 2000–1208 du 13 décembre 2000 relative à la solidarité et au renouvellement urbains.

47. Tonino Serafini, "Une ville étranglée par la crise du logement," *Libération*, 11 September 2007.

48. Amélia Blanchot, "Pour une chambre de 12 m2, 40 candidats dans l'escalier," *Libération*, 20 September 2007; Keren Lentschner, "Avis de pénurie sur le marché des chambres de bonne," *Le Figaro*, 9 April 2007; Didier Porquery, "Spéculations," *Libération*, 20 September 2007; Tonino Serafini, "Les mal-logés investissent la rue de la Banque," *Libération*, 5 October 2007.

49. But as the media have also reported, the price per square meter of these apartments is among the most exorbitant in Paris, and many of them do not even meet the minimum code requirements for rental. See Lentschner, "Avis de pénurie."

50. Laure Equy, "Sous la douche, je pense à l'endroit où je vais dormir," *Libération*, 22 January 2008.

51. Marcel Bresard, "Expose sur l'État d'avancement de l'enquête sur les conditions de vie des personnes âgées," n.d. [1960 or 1961], 10, Centre des Archives Contemporaines, Fontainebleau (henceforth CAC), 19860269–004.

52. Simone de Beauvoir, *La vieillesse* (Paris: Gallimard, 1970), 254–55. See also Catherine Bonvalet and Jim Ogg, "The Housing Situation and Residential Strategies of Older People in France," *Ageing & Society* 28 (2008): 753–77.

53. See, e.g., the archives of the Laboratoire d'Analyse Secondaire et des Méthodes Appliquées à la Sociologie (LASMAS) at the Institut de Recherche sur les Sociétés Contemporaines (IRESCO), Paris.

54. Those with psychiatric diagnoses were five times likelier to die during the heat wave than those without; see Institut de Veille Sanitaire, *Étude*, 60.

55. Interview with Danièle Alet, 2011; see also the profile of Patricia in Alet's film *Aux oubliés de la canicule.*

56. See *Les oubliés de la canicule*; also Chamak, Garcia, and Prieur, "Ces vieillards morts dans l'oubli."

57. Serafini, "Les mal-logés"; Serafini, "Encore à la rue," *Libération,* 1 November 2007.

58. Louis Henry and Maurice Febvay, "La situation du logement dans la région parisienne," *Population* 12, no. 1 (1957): 129–40; Yankel Fijalkow, "Surpopulation ou insalubrité: Deux statistiques pour décrire l'habitat populaire (1880–1914)," *Le mouvement social* 182 (1998): 79–96.

59. Ross, "Schoolteachers"; and Coffman, "The Papin Enigma."

60. Karl Laske, "Un enfant sans papiers fuit la police et chute du 4e étage," *Libération,* 10 August 2007; Laske, "Le saut dans le désespoir des sans papiers traqués," *Libération,* 21 September 2007.

61. See Jackson, *Paris under Water,* or Michael Eric Dyson, *Come Hell or High Water: Hurricane Katrina and the Color of Disaster* (New York: Basic, 2006), for example; also Ronald J. Daniels, Donald F. Kettl, and Howard Kunreuther, *On Risk and Disaster: Lessons from Hurricane Katrina* (Philadelphia: University of Pennsylvania Press, 2006). On hurricanes, see Steinberg, *Acts of God.*

62. Steinberg, "The Secret History of Natural Disaster"; Davis, *Ecology of Fear.* See also Erikson, *A New Species of Trouble*; and Fortun, *Advocacy after Bhopal.*

63. Cutter, Boruff, and Shirley, "Social Vulnerability to Environmental Hazards."

64. Giovanna Di Chiro, "Nature as Community: The Convergence of Environment and Social Justice," in *Uncommon Ground: Toward Reinventing Nature,* ed. William Cronon (New York: W. W. Norton, 1995), 298–320.

65. Jonas Ebbeson, ed., *Access to Justice in Environmental Matters in the EU* (The Hague: Kluwer Law International, 2002); Iris Marion Young, *Inclusion and Democracy* (New York: Oxford University Press, 2000).

Chapter Four

1. Chamak, Garcia, and Prieur, "Ces vieillards morts dans l'oubli"; and interviews with local residents, June and July 2007.

2. Anna Gavalda, *Ensemble, c'est tout* (Paris: Le dilettante, 2004).

3. Ibid., 10.

4. Ibid., 518. Gavalda writes that Camille is *accablée* by the dealer's words: literally, "overcome." But she engages in an interesting wordplay here, as journalists frequently described the heat's elderly victims as *accablés par la chaleur* or "overcome by the heat," and described the heat as *accablante* or "overwhelming."

5. Michel Houellebecq, *La possibilité d'une île* (Paris: Fayard, 2005), 90.

6. Ibid., 20.

7. Ibid., 385.

8. David Troyansky, *Old Age in the Old Regime: Image and Experience in Eighteenth-Century France* (Ithaca, NY: Cornell University Press, 1989); Patrice Bourdelais, *L'âge de la vieillesse: Histoire du vieillissement de la population* (Paris: Odile Jacob, 1993). See also Sherri Klassen, "Greying in the Cloister: The Ursuline Life Course in Eighteenth-Century France," *Journal of Women's History* 12, no. 4 (2001): 87–112.

9. See, e.g., Robert Nye, *Crime, Madness, and Politics in Modern France: The Medical Concept*

of National Decline (Princeton, NJ: Princeton University Press, 1984); also Daniel Pick, *Faces of Degeneration: A European Disorder* (New York: Cambridge University Press, 1993).

10. See Bourdelais, *L'âge de la vieillesse,* esp. 117–54; Mary Louise Roberts, *Civilization without Sexes: Reconstructing Gender in Postwar France, 1914–1939* (Chicago: University of Chicago Press, 1994); and Cheryl Koos, "Gender, Anti-Individualism, and Nationalism: The Alliance Nationale and the Pronatalist Backlash against the *Femme moderne,* 1933–1940," *French Historical Studies* 19, no. 3 (1996): 699–723.

11. See Fernand Boverat, *Comment nous vaincrons la dénatalité* (Paris: L'Alliance nationale contre la dépopulation, 1939). For more on Boverat, see Koos, "Gender."

12. See Alain Parant, "Croissance démographique et vieillissement," *Population* 47, no. 6 (1992): 1657–76; citation on 1658.

13. Jean Daric, *Vieillissement de la population et prolongation de la vie active,* Travaux et Documents, Institut National d'Études Démographiques, Cahier no. 7 (Paris: Presses Universitaires de France, 1948), 42.

14. Cited in Elise Feller, "Les femmes et le vieillissement dans la France du premier XXe siècle," *Clio* 7 (1998): 199–222; citation on 201; also see Feller, *Histoire de la vieillesse en France, 1900–1960: Du vieillard au retraité* (Paris: Seli Arslan, 2005).

15. On Mauco, see Patrick Weil, "Georges Mauco, expert en immigration: Ethnoracisme pratique et antisémitisme fielleux," in *L'antisémitisme de plume 1940–1944: Études et documents,* ed. Pierre-André Taguieff (Paris: Berg, 1999), 267–76.

16. Boverat to Mauco, 4 May 1949, CAC, 19860269–001.

17. Boverat, "Le vieillissement de la population et ses repercussions sur la sécurité sociale," Haut Comité Consultatif de la Population et de la Famille, 30 October 1951, 1–3 ter., CAC, 19860269–001.

18. Ibid., 3 ter.

19. Ibid., 6.

20. Ibid., 11.

21. Rémi Lenoir, "L'invention du 'troisième âge'," *Actes de la recherche en sciences sociales* 26–27 (1979): 57–82.

22. Karl Marx, *Capital: A Critique of Political Economy,* trans. Ben Fowkes (New York: Penguin, 1976), vol. 1, p. 518. Also Lenoir, "L'invention," 60–61.

23. Elise Feller, "L'entrée en politique d'un groupe d'âge: La lutte des pensionnés de l'État dans l'entre-deux-guerres et la construction d'un 'modèle français' de retraite," *Le mouvement social* 190 (2000): 33–59; also Dominique Argoud and Anne-Marie Guillemard, "The Politics of Old Age in France," in *The Politics of Old Age in Europe,* ed. Alan Walker and Gerhard Naegele (Buckingham: Open University Press, 1999), 83–92.

24. See, e.g., Peter Stearns, *Old Age in European Society: The Case of France* (London: Croom Helm, 1977), 58. Also Anne-Marie Guillemard, *Le déclin du social: Formation et crise des politiques de la vieillesse* (Paris: Presses Universitaires de France, 1986); and Bruno Dumons and Gilles Pollet, "La question des retraites vue par les socialistes français (1880–1956)," in *Regards croisés sur la protection sociale de la vieillesse,* ed. Françoise Cribier and Elise Feller (Paris: Comité d'histoire de la sécurité sociale, 2005), 197–224.

25. See especially Dumons and Pollet, "La question des retraites."

26. *Journal Officiel,* 31 March 1910.

27. Guillemard, *Le déclin du social.*

28. Christel Chaineaud, "L'État et la veuve âgée sous la IIIe République: Entre indifference et prise en charge," in *De l'hospice au domicile collectif: La vieillesse et ses prises en charge de la fin*

du XVIIIe siècle à nos jours, ed. Yannick Marec and Daniel Réguer (Mont-Saint-Aignan: Presses Universitaires de Rouen et du Havre, 2013), 419–27.

29. Paul Dutton, *Origins of the French Welfare State: The Struggle for Social Reform in France, 1914–1947* (New York: Cambridge University Press, 2002). See also Susan Pedersen, *Family, Dependence, and the Origins of the Welfare State: Britain and France, 1914–45* (New York: Cambridge University Press, 1993).

30. Bourdelais, *L'âge de la vieillesse*, 77.

31. Feller, *Histoire de la vieillesse en France.*

32. Vincent Caradec, "'Seniors' et 'personnes âgées': Réflexions sur les modes de catégorisation de la vieillesse," in Cribier and Feller, *Regards croisés sur la protection sociale de la vieillesse*, 313–26; citation on 313; emphasis in the original.

33. Ibid., 314; and Isabelle Mallon, *Vivre en maison de retraite: Le dernier chez-soi* (Rennes: Presse Universitaire de Rennes, 2005). As Mallon argues, there is a powerful contrast between images of new retirees enjoying their grandchildren and dedicating themselves to charity work, consumption, and leisure, and the very old, living in retirement homes or hospitals: representations thus "oscillate between admiration and pity" (7).

34. See in particular the discussion of the distinctions between discipline and security in Michel Foucault, *Security, Territory, Population: Lectures at the Collège de France, 1977–1978*, trans. Graham Burchell (New York: Picador, 2007), 44–46.

35. Ibid., 70–71.

36. Foucault, *"Society Must Be Defended": Lectures at the Collège de France, 1975–1976*, trans. David Macey (New York: Picador, 2003), 244.

37. Ibid.

38. Boverat, "Le vieillissement," 22–24.

39. Boverat, "Projet de motion concernant le vieillissement de la population," 14 November 1951, CAC, 19860269–001.

40. Foucault, *The History of Sexuality*, vol. 1, pp. 136, 138.

41. The term originates with Étienne Balibar, *Politics and the Other Scene*, trans. Christine Jones, James Swenson, and Chris Turner (London: Verso, 2002), 38 n. 33; but see especially Roberto Esposito, *Bíos: Biopolitics and Philosophy*, trans. Timothy Campbell (Minneapolis: University of Minnesota Press, 2008).

42. Jacques Bertillon, "De la population de la France et des remèdes à y apporter," *Journal de la société statistique de Paris* 36 (1895): 410–38; citation on 432.

43. Ibid., 433. This quotation, as well as the one above, also appears in Bourdelais, *L'Age de la vieillesse*, 93–94.

44. Boverat, "Le vieillissement," 7.

45. Ibid., 5.

46. Ibid., 21.

47. HCPF, *Politique de la vieillesse*, 3 (henceforth "Rapport Laroque").

48. Ibid., 3–4.

49. Ibid., 19.

50. Ibid., 6.

51. Adriana Petryna, *Life Exposed: Biological Citizens after Chernobyl* (Princeton, NJ: Princeton University Press, 2002).

52. For another analysis of how grassroots political participation can avoid these binds, see Young, *Inclusion and Democracy*.

53. Rapport Laroque, 9.

54. Preamble to the Charter of the United Nations and Statute of the International Court of Justice, 26 June 1945; UN General Assembly, *Universal Declaration of Human Rights*, 10 December 1948; and Preamble to the Constitution of the World Health Organization as adopted by the International Health Conference, New York, 19 June–22 July 1946. Such an imagination was pervasive in French family policy: see Camille Robcis, *The Law of Kinship: Anthropology, Psychoanalysis, and the Family in France* (Ithaca, NY: Cornell University Press, 2013).

55. IFOP poll, 82, CAC, 19860269–006.

56. Ibid., 93.

57. Ibid.

58. Association Nationale des Assistantes Sociales et des Assistants Sociaux, "Étude sur les reactions et attitutes individuelles des personnes âgées en face des problèmes de la vieillesse et sur les difficultés que rencontrent les assistantes sociales dans leur action auprès des personnes âgées," March 1961, 8–9, CAC, 19860269–004.

59. Rapport Laroque, 41–45.

60. Procès-verbaux de la réunion du 8 Octobre 1960 de la Commission d'Études du Problème de la Vieillesse," 9, CAC, 19860269–006.

61. Association Nationale des Assistantes Sociales et des Assistants Sociaux, "Étude," 16, 18.

62. IFOP poll, 93.

63. Association Nationale des Assistantes Sociales et des Assistants Sociaux, "Étude," 2, 16.

64. Boverat to Georges Mauco, 4 February 1962, CAC, 19860269–001.

65. See, e.g., Henri Amouroux, *1945–1950: La France du baby boom* (Paris: La Découverte, 1991); Yvonne Kniebiehler, *La révolution maternelle: Femmes, modernité, citoyenneté depuis 1945* (Paris: Perrin, 1997); Louis Chauvel, *Le destin des générations: Structure sociale et cohortes en France au XXe siècle* (Paris: Presses Universitaires de France, 1998); Jean-François Sirinelli, *Les baby-boomers: Une génération, 1945–1969* (Paris: Fayard, 2003).

66. See, e.g., Dutton, *Origins of the French Welfare State*; Timothy B. Smith, *France in Crisis: Welfare, Inequality, and Globalization since 1980* (New York: Cambridge University Press, 2004).

67. Stearns, *Old Age in European Society*, 18.

68. Rapport Laroque, 259.

69. Georges Mauco, "L'apport humain et social des personnes âgées," *La Liberté*, 13 May 1963.

70. "Les personnes âgées ont-elles oui ou non le droit de vivre?," *France-Soir*, 20 June 1962.

71. "Une vieillesse sacrifiée," *France-Soir*, 1 March 1962.

72. Castel, *Les métamorphoses.*

73. Association Nationale des Assistantes Sociales et des Assistants Sociaux, "Étude," 16.

74. See Castel, *Les métamorphoses*, 295.

75. On the rise of consumerism and its relationship to national rejuvenation, see Kristin Ross, *Fast Cars, Clean Bodies: Decolonization and the Reordering of French Culture* (Cambridge, MA: MIT Press, 1996); and Richard Kuisel, *Seducing the French: The Dilemma of Americanization* (Berkeley: University of California Press, 1993). See also Victoria de Grazia, *Irresistible Empire: America's Advance through 20th-Century Europe* (Cambridge, MA: Harvard University Press, 2005).

76. Rapport Laroque, 259.

77. See especially Richard I. Jobs, *Riding the New Wave: Youth and the Rejuvenation of France after the Second World War* (Stanford, CA: Stanford University Press, 2007).

78. Mauco, "L'apport humain."

79. Lenoir, "L'invention," 66. See also Feller, *Histoire de la vieillesse*; Guillemard, *Le déclin social.*

80. See especially Agamben, *Homo Sacer*; Agamben, *State of Exception*, trans. Kevin Attell (Chicago: University of Chicago Press, 2005); also Michael Hardt and Antonio Negri, *Empire* (Cambridge, MA: Harvard University Press, 2000), 366.

81. Loi n°70–1318 du 31 décembre 1970 portant réforme hospitalière; Loi n° 71–582 du 16 juillet 1971 relative à l'allocation de logement.

82. Beauvoir, *La vieillesse*, 231.

83. Ibid., 240.

84. "Quand on n'est plus capable de faire un travailleur, on est tout juste bon à faire un macchabée"; Beauvoir, *La vieillesse*, 257–58.

85. Beauvoir, *La vieillesse*, 292.

86. Lenoir, "L'invention," 64.

87. Nicole Benoît-Lapierre, Rithée Cevasco, and Markos Zafiropoulos, *Vieillesse des pauvres: Les chemins de l'hospice* (Paris: Éditions Économie et Humanisme, 1980).

88. André Champigny, "Les dix commandements de la vieillesse," *Gérontologie* 8 (1972): 36.

89. Lenoir, "L'invention," and Jacqueline Trincaz, "Les fondements imaginaires de la vieillesse dans la pensée occidentale," *L'homme* 38 (1999): 167–89, cite them as nonsatirical.

90. Most of the documents I cite here are located in the archives of the Laboratoire d'Analyse Secondaire et de Méthodes Appliquées à la Sociologie (LASMAS) at the Centre Maurice Halbwachs, Paris. There are several collections of documents relating to these cohorts, including the interview series "Jeunes provinciaux d'hier, vieux Parisiens d'aujourd'hui," conducted by the Équipe de Géographie Sociale et Gérontologie. The collection is generally well classified, with most documentation organized by individual. I will cite, where available, the archive's code for the individual; the group has changed the names of all interviewees for confidentiality purposes in the coding process.

91. One of the chief questions interviewers pursued was, "Do you feel more like a Parisian or like someone from the provinces who lives in Paris?" Other questions included the following: "What are Parisians like? How are they different from the people from [interviewee's region]?"

92. At the project's beginning, most subjects had only recently retired and had relatively little to say about growing old; by the 1990s, many of the participants had died.

93. LASMAS GU05; ellipses in original, indicating a pause, not an excising of text.

94. LASMAS LE44.

95. LASMAS "Delettraz." Some cases are not filed by code, and are instead classified by (fictional) surname.

96. LASMAS "Lerver."

97. LASMAS "Vandamele."

98. LASMAS LE12.

99. LASMAS "Berty."

100. LASMAS KE04.

101. LASMAS TR06.

102. LASMAS BI06.

103. LASMAS "Guitard."

104. LASMAS GU05.

105. LASMAS "Leroux."

106. LASMAS BI06.

107. LASMAS PL04.

108. LASMAS "Rabeau."

109. Ibid.

110. LASMAS CA08.

111. LASMAS BI06.

112. LASMAS CO24.

113. LASMAS "Bonnard."

114. LASMAS "Brasc"; AR03.

115. LASMAS KE04.

116. LASMAS MA51; "Castaing."

117. LASMAS "Poussot."

118. LASMAS BE39; ellipses in original.

119. LASMAS "Giron."

120. LASMAS "Delattraz."

121. LASMAS "Guitard."

122. LASMAS AR03.

123. LASMAS "Castaing."

124. LASMAS BA16.

125. LASMAS LE12.

126. LASMAS AU06.

127. Françoise Cribier, "Vieillesse et citoyenneté," in *Villes et vieillir* (Paris: La Documentation française, 2004), 312–19; citation on 314. The paper was delivered in January 2003, and was in press before the heat wave struck.

128. LASMAS RO14.

129. This is a somewhat misleading figure. Some 40 percent of those who died during the heat wave died in hospitals, while some 20 percent died in nursing homes. Yet some of those who died in hospitals had been transferred from nursing homes. There are no data on how many who died during the heat wave *lived* in nursing homes but did not die there.

130. Josiane Holstein et al., "Were Less Disabled Patients the Most Affected by 2003 Heat Wave in Nursing Homes in Paris, France?," *Journal of Public Health* 27, no. 4 (2005): 359–65; interview with Alfred Spira, June 2005.

131. Fascicule spécial n° 65–38 bis, "Normes des maisons de retraite."

132. Beauvoir, *La vieillesse*, 271.

133. France 3, *Les dossiers de France 3: Vieillir ensemble?*, 23 September 2003.

134. Jean-François Lacan, *Scandales dans les maisons de retraite* (Paris: Albin Michel, 2002); also Groupe Épidémiologie du Département Action Scientifique en Médecine du Travail du Centre Interservices de Santé et de Médecine du Travail en Entreprise, *Le travail d'aide aux personnes âgées: Conditions de travail et santé perçue chez les aides soignants, agents de service, aides ménagères des maisons de retraite et des structures d'aide à domicile*, Publications ASMT 17 (Paris: Éditions Docis, 1999); and Mallon, *Vivre en maison de retraite*.

135. Gerard Badou, *Les nouveaux vieux* (Paris: Le Pré aux clercs, 1989), 249.

136. João Biehl, *Vita: Life in a Zone of Social Abandonment* (Berkeley: University of California Press, 2005).

137. See Smith, *France in Crisis*, for a useful account.

138. Badou, *Les nouveaux vieux*.

139. "Le comique de Bercy," *Le Canard Enchaîné*, 6 August 2003.

140. "Francis Mer, devin d'honneur," *Le Canard Enchaîné*, 20 August 2003.

141. Figures from Insee, 2008.

142. Figures from Insee, 2006.

143. Figures from Insee, 2013.

144. Céline Arnold and Michèle Lelièvre, "Le niveau de vie des personnes âgées de 1996 à 2009: Une progression moyenne en ligne avec celle des personnes d'âge actif, mais des situations individuelles et générationnelles plus contrastées," in *Les revenus et le patrimoine des ménages* (Paris: Insee, 2013), 33–53.

145. France 3, *Vieillir ensemble.*

146. See Jean-François Sirinelli, "Appel des intellectuels en soutien aux grévistes," *Libération*, 12 January 1998. For Bourdieu's speech, see Jill Forbes, "Bourdieu's Maieutics," *SubStance* 29, no. 3 (2000): 22–42.

147. Timothy B. Smith, *France in Crisis: Welfare, Inequality and Globalization since 1980* (New York: Cambridge University Press, 2004), 154, 158.

148. I thank Dana Simmons for pointing this out.

149. I thank Dana Simmons for this phrasing.

Chapter Five

1. Interview, 11 June 2007.

2. Subsequent research revealed that the correct Suzanne had resided in the Paris *banlieue*, and was therefore outside the scope of the project.

3. "Catastrophic Modeling and California Earthquake Risk: A 20-Year Perspective," RMS Special Report (Risk Management Solutions, 2009), http://www.rms.com/publications/LomaPrieta_20Years.pdf; Morgan O'Rourke, "The Property/Casualty Market in 2011," *Risk Management*, February 2011.

4. Mortality estimates can range wildly for the same disaster. For example, estimates for the Haitian earthquake range from 316,000 (the state's official figure) to between 46,000 and 85,000 (according to an unpublished USAID report: see Timothy T. Schwartz, *Building Assessments and Rubble Removal in Quake-Affected Neighborhoods in Haiti: BARR Survey Final Report* [Washington, DC: USAID, 2011]). The state has provided no methodological disclosures for its high estimate. For much of 2010, the government assessed the death toll at 230,000, and it abruptly raised that figure on the anniversary of the earthquake with no explanation. The USAID report, by contrast, came to its mortality estimate indirectly. It is based on a random survey of damage and deaths in selected households in Port-au-Prince, which it then extrapolated to the general population.

5. See Petryna, *Life Exposed.*

6. James Jurin, "A Letter . . . Containing a Comparison between the Danger of the Natural Small Pox, and of That Given by Inoculation," *Philosophical Transactions* 32 (1722–23): 213–27.

7. See Harry Marks, "When the State Counts Lives: Eighteenth-Century Quarrels over Inoculation," in *Body Counts: Medical Quantification in Historical and Sociological Perspective/La quantification médicale, perspectives historiques et sociologiques*, ed. Gérard Jorland, Annick Opinel, and George Weisz (Montreal: McGill/Queen's University Press, 2005), 51–64.

8. Foucault, *"Society Must Be Defended,"* 243–48.

9. See Andrea Rusnock, *Vital Accounts: Quantifying Health and Population in Eighteenth-Century England and France* (New York: Cambridge University Press, 2002); also Cole, *The Power of Large Numbers*, 33–44.

10. Coleman, *Death Is a Social Disease*, 139. Andrea Rusnock challenges this view, arguing that hospital registers served as an important corrective to the baptism problem even in the late

eighteenth century; see Rusnock, "Quantifying Infant Mortality in England and France, 1750–1800," in Jorland, Opinel, and Weisz, *Body Counts*, 65–86.

11. See Bruno Latour, *The Pasteurization of France*, trans. Alan Sheridan and John Law (Cambridge, MA: Harvard University Press, 1988).

12. See Coleman, *Death Is a Social Disease*, 149–71; Cole, *The Power of Large Numbers*, 69; Barnes, *The Making of a Social Disease*, 32–33.

13. Cole, *The Power of Large Numbers*, 72; also Coleman, *Death Is a Social Disease*, 179.

14. See Theodore M. Porter, "Medical Quantification: Science, Regulation, and the State," in Jorland, Opinel, and Weisz, *Body Counts*, 394–401.

15. See Cole, *The Power of Large Numbers*, 100–101.

16. Cole, *The Power of Large Numbers*, 76–77.

17. Abenhaïm, *Canicules*, 91–112.

18. TF1, *20 heures*, 10 and 11 August 2003.

19. See Vassy, Keller, and Dingwall, *Enregistrer les morts*, for a deeper analysis of this problem.

20. See "L'État finit par entendre le SOS des médecins," *Libération*, 14 August 2003.

21. Rapport Jacquat, vol. 2, p. 109.

22. TF1, *20 heures*, 16 August 2003.

23. Cited in Abenhaïm, *Canicules*, 101.

24. Abenhaïm, *Canicules*, 96–101. Also, e.g., France 2, *20 heures—Le journal*, 18 August 2003.

25. See, e.g., France 2, *20 heures—Le journal*, 20 August 2003; TF1, *20 heures*, 20 August 2003.

26. Much of this account is based on extensive conversations with demographers at Inserm who produced the official report on mortality during the 2003 heat wave, which I conducted (along with the sociologist Carine Vassy at Inserm's CépiDc facility and at the École des Hautes Études en Sciences Sociales in Paris in May, June, and July 2007). For further discussion of mortality recording during the heat wave, see Vassy, Keller, and Dingwall, *Enregistrer les morts*.

27. Hémon and Jougla, *Surmortalité—Rapport*, 21.

28. "La canicule de 2006 a fait 133 morts," *Le Nouvel Observateur*, 24 October 2006. This number was itself an upward revision of the first official reports of 64 deaths (see France 2, *20 heures—Le journal*, 27 July 2006) and 112 deaths (see France 2, *20 heures—Le journal*, 3 August 2006).

29. Alain Le Tertre et al., "Première estimation de l'impact de la vague de chaleur sur la mortalité durant l'été 2006," *Bulletin épidémiologique hebdomadaire* 22–23 (2007): 190–92; Anne Fouillet, "Comparaison de la surmortalité observée en juillet 2006 à celle estimée à partir des étés 1975–2003, France," ibid., 192–97.

30. Le Tertre, "Première estimation," 191.

31. As Laurent Toulemon and Magali Barbieri, "The Mortality Impact of the August 2003 Heat Wave in France: Investigating the 'Harvesting Effect' and Other Long-Term Consequences," *Population Studies* 62, no. 1 (2008): 41, note, there is no standard in the field for measuring harvesting, with some considering displaced mortality over days, and others over months or years.

32. Indeed, most studies of heat waves have found little to no evidence of displaced mortality. See R. Basu and B. Malig, "High Ambient Temperature and Mortality in California: Exploring the Roles of Age, Disease, and Mortality Displacement," *Environmental Research* 111, no. 8 (2011): 1286–92; C. Huang et al., "Projecting Future Heat-Related Mortality under Climate Change Scenarios: A Systematic Review," *Environmental Health Perspectives* 119, no. 12 (2011): 681–90; Maud M. T. E. Huynen et al., "The Impact of Heat Waves and Cold Spells on

Mortality Rates in the Dutch Population," *Environmental Health Perspectives* 109, no. 5 (2001): 463–70. R. S. Kovats and S. Hajat, "Heat Stress and Public Health: A Critical Review," *Annual Review of Public Health* 29 (2008): 41–55, indicate how murky the data on harvesting effects can be. S. Hajat et al., "Mortality Displacement of Heat-Related Deaths: A Comparison of Delhi, São Paulo, and London," *Epidemiology* 16, no. 5 (2005): 613–20, found a slight harvesting effect in one of their cases but not the others. In their study based on temperature and mortality in twelve US cities, Alfésio Luís Ferreira Braga, Antonella Zanobetti, and Joel Schwartz, "The Time Course of Weather-Related Deaths," *Epidemiology* 12, no. 6 (2001): 662–67, argue that heat deaths are "primarily harvesting," without clarifying their definition.

33. Alain-Jacques Valleron and Ariane Boumendil, "Épidémiologie et canicules: Analyses de la vague de chaleur 2003 en France," *Comptes rendus: Biologies* 327 (2004): 1125–41.

34. Daniel Rousseau, "Surmortalité des étés caniculaires et surmortalité hivernale en France," *Climatologie* 3 (2006): 43–54.

35. Patrick Lagadec and Hervé Laroche, *Retour sur les rapports d'enquête et d'expertise suite à la canicule de l'été 2003* (Grenoble: Maison des Sciences de l'Homme-Alpes, 2005), 49.

36. Grégoire Rey, "Vagues de chaleur, fluctuations ordinaires des températures et mortalité en France depuis 1971," *Population* 62, no. 3 (2007): 533–63.

37. Toulemon and Barbieri, "The Mortality Impact."

38. Valleron and Boumendil, "Épidémiologie et canicules," 1137.

39. Létard et al., Rapport d'information fait au nom de la mission commune d'information, "La France et les Français face à la canicule: les leçons d'une crise," Sénat, no. 195, Session ordinaire de 2003–2004, p. 132.

40. See Patrick J. Michaels, ed., *Shattered Consensus: The True State of Global Warming* (Arlington: The George C. Marshall Institute, 2005), 190.

41. Étienne Israelewicz, "Effet moisson: L'impact des catastrophes vie sur la mortalité à long terme," *Bulletin français d'actuariat* 12, no. 24 (2012): 113–59.

42. Butler, *Precarious Life*, 20. See also Butler, *Frames of War*.

43. Butler, *Precarious Life*, 36.

44. See especially Scott, "French Universalism"; also Scott, *Parité! Sexual Equality and the Crisis of French Universalism* (Chicago: University of Chicago Press, 2005).

45. Julien Damon, "La citoyenneté du SDF" (DEA thesis, Université de Paris IV, 1994). See also Turner, "Outline of a Theory of Citizenship"; Loïc Wacquant, *Urban Outcasts: A Comparative Sociology of Advanced Marginality* (Cambridge: Polity Press, 2008); and Didier Fassin, *La force de l'ordre: Une anthropologie de la police des quartiers* (Paris: Seuil, 2011).

46. Mairie de Paris, Vingtième Arrondissement, Acte de décès no. 1025, 7 August 2003.

47. Mairie de Paris, Dixième Arrondissement, Acte de décès no. 1001, 12 August 2003.

48. See, for example, Maxime Blondet, "Les oubliés reposent en terre," *France-Soir*, 4 September 2003, 6; as well as the articles and news programs cited in chapter 2.

49. See Mitchell Duneier, "Ethnography, the Ecological Fallacy, and the 1995 Chicago Heat Wave," *American Sociological Review* 71 (2006): 679–88; and Eric Klinenberg, "Blaming the Victims: Hearsay, Labeling, and the Hazards of Quick-Hit Disaster Ethnography," ibid., 689–98.

50. Indeed, probability militates strongly in favor of these deaths as heat related, given that these deaths all occurred during the peak of the heat wave, whereas the figures for admission to the cemetery include deaths that occurred in the last ten days of August.

51. See Toulemon and Barbieri, "The Mortality Impact," 42–44.

52. For Noël, see Mairie de Paris, Dixième Arrondissement, Acte de décès no. 986, 11 August 2003.

53. Julie Lasterade, "Noyades accidentelles: + 39 percent pendant l'été," *Libération*, 23 September 2003.

54. This does not have to be the case, at least when disaster strikes on a smaller scale. In the case of the Chicago heat wave of 1995, for example, Cook County medical examiner Edmund R. Donoghue took account of a range of factors in his attribution of cause of death, including the living conditions of the decedent and other circumstantial evidence, to widen the category of heat death. As a consequence official estimates of cause of death put the toll at 465, while an excess death study put the figure at 739: the excess death figure is still greater, but the difference is far smaller than that produced in the French heat wave of 2006. Indeed, Donoghue told the press, "If anything, we're underestimating the amount of death." See Klinenberg, *Heat Wave*, 28–29. But given the scale of the disaster in France in 2003, such a detailed forensic analysis of so many affected communities would be unfeasible.

55. Hémon and Jougla, *Surmortalité—Rapport*, 22.

56. Ibid., 30.

57. Ibid., 32.

58. Ibid., 30.

59. Lettre circulaire du 10 février 2004 du secrétaire d'État aux personnes âgées relative à la prévention des conséquences d'une nouvelle période de canicule dans les établissements d'hébergement pour personnes âgées (EHPA).

60. See Katherine M. Shea, "Global Climate Change and Children's Health," *Pediatrics* 120 (2007): 1359–67; Xavier Basagaña et al., "Heat Waves and Cause-Specific Mortality at All Ages," *Epidemiology* 22, no. 6 (2011): 765–72; Thomas Waters, "Heat Illness: Tips for Recognition and Treatment," *Cleveland Clinic Journal of Medicine* 68, no. 8 (2001): 685–87.

61. Denis Hémon et al., "Surmortalité liée à la canicule d'août 2003 en France," *Bulletin épidémiologique hebdomadaire* 45–46 (2003): 221–25.

62. Institut de Veille Sanitaire, *Impact sanitaire de la vague de chaleur d'août 2003 en France: Bilan et perspectives—Octobre 2003* (Paris: InVS, 2003), 55, 26.

63. See Jean-Pierre Besancenot, "Vagues de chaleur et mortalité dans les grandes agglomérations urbaines," *Environnement, risques et santé* 1, no. 4 (2002): 229–40, which refers to elevated child mortality among migrant Roma populations during heat waves in particular.

64. InVS, *Impact sanitaire*, 26, 55.

65. Interview, June 2005.

66. Rivaled only by the western half of the seventeenth arrondissement. See Canouï-Poitrine, Cadot, and Spira, "Excess Deaths"; see also Cadot and Spira, "Canicule et surmortalité," 242–43.

67. Grégoire Rey et al., "Heat Exposure and Socio-Economic Vulnerability as Synergistic Factors in Heat-Wave-Related Mortality," *European Journal of Epidemiology* 24 (2009): 495–502. This is according to a deprivation index that "analyse[s] socioeconomic spatial mortality differentials" in France by "urban unit category," or neighborhood (496).

68. Joël Belmin et al., "Level of Dependency: A Simple Marker Associated with Mortality during the 2003 Heatwave among French Dependent Elderly People Living in the Community or in Institutions," *Age and Ageing* 36 (2007): 298–303.

69. Hémon and Jougla, *Surmortalité—Rapport*, 43; interview with Alfred Spira, Bicêtre, June 2005.

70. Hémon and Jougla, *Surmortalité—Rapport*, 55.

71. Lynette Cusack, Charlotte de Crespigny, and Peter Athanasos, "Heatwaves and Their Impact on People with Alcohol, Drug, and Mental Health Conditions: A Discussion Paper on

Clinical Practice Considerations," *Journal of Advanced Nursing* 67, no. 4 (2011): 915–22; Bo Li et al., "The Impact of Extreme Heat on Morbidity in Milwaukee, Wisconsin," *Climatic Change* 110, nos. 3–4 (2011): 959–76; Jan C. Semenza et al., "Heat-Related Deaths during the July 1995 Heat Wave in Chicago," *New England Journal of Medicine* 335 (1996): 84–90; and Klinenberg, *Heat Wave.*

72. "Kaposi's Sarcoma and *Pneumocystis* Pneumonia among Homosexual Men—New York and California," *Morbidity and Mortality Weekly Report* 30, no. 25 (1981): 305–7. Also "*Pneumocystis* Pneumonia—Los Angeles," *Morbidity and Mortality Weekly Report* 30, no. 21 (1981): 250–52.

73. See Steven Epstein, *Impure Science: AIDS, Activism, and the Politics of Knowledge* (Berkeley: University of California Press, 1996); Paula Treichler, *How to Have Theory in an Epidemic: Cultural Chronicles of AIDS* (Durham, NC: Duke University Press, 1999); Paul Farmer, *Infections and Inequalities: The Modern Plagues* (Berkeley: University of California Press, 1999). For examples of contemporary denialism of heterosexual AIDS, see Michael Fumento, *The Myth of Heterosexual AIDS: How a Tragedy Has Been Distorted by the Media and Partisan Politics* (New York: Regnery, 1993); Elizabeth Pisani, *The Wisdom of Whores: Bureaucrats, Brothels, and the Business of AIDS* (New York: Norton, 2009); and Michelle Cochrane, *When AIDS Began: San Francisco and the Making of an Epidemic* (New York: Routledge, 2003).

Epilogue

1. Thierry Jonquet, *Mon vieux* (Paris: Seuil, 2004).

2. Ibid., 236–37.

3. Ibid., 233.

4. Michael K. Gusmano, Victor Rodwin, and Alfred Spira, "Aging in New York City and Paris: Challenges for Research and Policy" (public presentation, Lectures in Public Policy Series, American University of Paris, 5 June 2007).

5. Mairie de Paris, *Référentiel: Un aménagement durable pour Paris,* 5th ed. (Paris, 2010), 28, 70.

6. See, e.g., "Eden Bio by Édouard François," *De Zeen,* http://www.dezeen.com/2009/02/17/eden-bio-by-edouard-francois-2/; Alexandra Kain, "Eden Bio: Paris Grows a Green Heart," *Inhabitat,* 5 May 2009, http://inhabitat.com/eden-bio/; "Eden Bio," *Smart Urbanism,* http://www.smarturb.org/index.php?#article:93; and Terri Peters, "Housing on Rue des Vignoles," *Architecture Week,* 17 March 2010, http://www.ArchitectureWeek.com/2010/0317/index.html.

7. Nancy Rose Hunt, "An Acoustic Register, Tenacious Images, and Congolese Scenes of Rape and Repetition," *Cultural Anthropology* 23, no. 2 (2008): 220–53.

8. Although I wrote cemetery officials to try to determine why they were still there after so long, I received no reply.

Bibliography

Archives

Centre des Archives Contemporaines, Fontainebleau.

Institut National de l'Audiovisuel, Paris.

Laboratoire d'Analyse Secondaire et des Méthodes Appliquées à la Sociologie (LASMAS), Institut de Recherche sur les Sociétés Contemporaines (IRESCO), Paris.

Films

Aux oubliés de la canicule. Directed by Danièle Alet. France, 2004.

Chambre de bonne. Directed by Maija-Lene Rettig. Germany, 2002.

Les femmes du 6e étage. Directed by Philippe Le Guay. France, 2011.

Les oubliés de la canicule. Directed by Sophie Lepault and Ibar Aibar. France, 2003–4.

Reports

Assemblée Nationale. *Débats parlementaires: Compte rendu integral des séances du mardi 7 octobre 2003.* 8 October 2003. http://www.assemblee-nationale.fr.

———. No. 1091. *Rapport d'information sur la crise sanitaire et sociale déclenchée par la canicule.* 2 vols. 24 September 2003. http://www.assemblee-nationale.fr.

———. No. 1455. *Rapport fait au nom de la Commission d'enquête sur les conséquences sanitaires et sociale de la canicule.* 2 vols. 25 February 2004. http://assemblee-nationale.fr.

Bastianelli, Jean-Paul, et al. *Mission d'enquête sur les fermetures de lits en milieu hospitalier durant l'été 2003.* Report no. 2003141. Paris: IGAS, 2003.

European Environment Agency. *Impacts of Europe's Changing Climate: An Indicator-Based Assessment.* Issue 2. Luxembourg: Office for Official Publications of the European Communities, 2004.

Fouillet, Anne, Grégoire Rey, Eric Jougla, and Denis Hémon. *Estimation de la surmortalité observée et attendue au cours de la vague de chaleur du mois de juillet 2006.* Paris: Inserm, 2006.

Groupe Épidémiologie du Département Action Scientifique en Médecine du Travail du Centre Interservices de Santé et de Médecine du Travail en Entreprise. *Le travail d'aide aux personnes âgées: Conditions de travail et santé perçue chez les aides soignants, agents de service,*

aides ménagères des maisons de retraite et des structures d'aide à domicile. Publications ASMT 17. Paris: Éditions Docis, 1999.
Haut Comité Consultatif de la Population et de la Famille. *Politique de la vieillesse: Rapport de la commission d'étude des problèmes de la vieillesse*. Paris: La Documentation française, 1962.
Hémon, Denis, and Eric Jougla. *Surmortalité liée à la canicule d'août 2003—Rapport d'étape: Estimation de la surmortalité et principales caractéristiques épidémiologiques*. Paris: Inserm, 2003.
———. *Surmortalité liée à la canicule d'août 2003: Suivi de la mortalité (21 août—31 décembre 2003); Causes médicales des décès (1–20 août 2003)*. Paris: Inserm, 2004.
Institut de Veille Sanitaire. *Étude des facteurs de risque de décès des personnes âgées résidant à domicile durant la vague de chaleur d'août 2003*. Paris: InVS, 2004.
———. *Impact sanitaire de la vague de chaleur d'août 2003 en France: Bilan et perspectives—Octobre 2003*. Paris: InVS, 2003.
Lalande, Françoise, et al. *Mission d'expertise et d'évaluation du système de santé pendant la canicule de 2003*. Paris: Ministère de la santé, de la famille et des personnes handicapées, 2003.
Létard, Valérie, et al. *Rapport au Sénat: La France et les Français face à la canicule*. 2004. www.senat.fr.

Newspapers

Aujourd'hui en France
Le Canard Enchainé
La Croix
La Dépêche
L'Express
Le Figaro
France-Soir
La Liberté
Le Monde
Le Parisien

Published Sources

"450 morts oubliés: Tous coupables." *Le Parisien*, 26 August 2003.
Abenhaïm, Lucien. *Canicules: La santé publique en question*. Paris: Fayard, 2003.
Abouchan, Khadija. "Chaleur fatale à l'hôpital." *Libération*, 12 August 2003.
Adveev, Alexandre, et al. "Populations et tendances démographiques des pays européens (1980–2010)." Population-F 66, no. 1 (2011): 9–133.
Agamben, Giorgio. *Homo Sacer: Sovereign Power and Bare Life*. Translated by Daniel Heller-Roazen. Stanford, CA: Stanford University Press, 1998.
———. *State of Exception*. Translated by Kevin Attell. Chicago: University of Chicago Press, 2005.
Allain, Pierre-Henri. "Canicule: adieu poulet, dindon, cochon . . ." *Libération*, 11 August 2003.
Alter, Anna, and Perrine Cherchève. "Mihajlo Molerovic, radié de la vie." *Marianne* 334 (15–21 September 2003): 24–25.
"L'amertume d'Hubert Falco." *Libération*, 21 August 2003.
Amouroux, Henri. *1945–1950: La France du baby boom*. Paris: La Découverte, 1991.

Argoud, Dominique, and Anne-Marie Guillemard. "The Politics of Old Age in France." In *The Politics of Old Age in Europe*, edited by Alan Walker and Gerhard Naegele, 83–92. Buckingham: Open University Press, 1999.

Arnaud, Didier. "À ses morts oubliés, la patrie repentante." *Libération*, 4 September 2003.

———. "Canicule: 14,802 morts." *Libération*, 26 September 2003.

Arnold, Céline, and Michèle Lelièvre. "Le niveau de vie des personnes âgées de 1996 à 2009: Une progression moyenne en ligne avec celle des personnes d'âge actif, mais des situations individuelles et générationnelles plus contrastées." In *Les revenus et le patrimoine des ménages*, 33–53. Paris: Insee, 2013.

"À ses morts oubliés, la patrie repentante." *Libération*, 4 September 2003.

Aubry, Olivier. "Les 35 heures ont aggravé les tensions dans les hôpitaux." *Le Parisien*, 14 August 2003.

Auffray, Alain. "Sarkozy se croyait sauvé." *Libération*, 25 September 2003.

Auster, Paul. *The Invention of Solitude*. New York: Penguin, 1982.

Authier, Danièle. "Représentations sociales du sentiment d'insécurité chez des personnes âgées résidant dans des quartiers opposés par leur taux de criminalité." MD thesis, Institut Universitaire Alexandre Lacassagne, Lyon, 1986.

Badou, Gerard. *Les nouveaux vieux*. Paris: Le Pré aux clercs, 1989.

Balibar, Étienne. *Politics and the Other Scene*. Translated by Christine Jones, James Swenson, and Chris Turner. London: Verso, 2002.

Bambaradeniya, Channa N. B., et al. *A Report on the Terrestrial Assessment of Tsunami Impacts on the Coastal Environment in Rekawa, Ussangoda and Kalametiya (RUK) Area of Southern Sri Lanka*. Colombo, Sri Lanka: IUCN-World Conservation Union Sri Lanka Country Office, 2005.

Baranowski, Shelley. *Strength through Joy: Consumerism and Mass Tourism in the Third Reich*. New York: Cambridge University Press, 2007.

"Barbarie française." *Le Figaro*, 25 August 2003.

Barnes, David. *The Great Stink of Paris and the Nineteenth-Century Struggle against Filth and Germs*. Baltimore: Johns Hopkins University Press, 2006.

———. *The Making of a Social Disease: Tuberculosis in Nineteenth-Century France*. Berkeley: University of California Press, 1995.

Barou, Jacques. *La place du pauvre: Histoire et géographie de l'habitat HLM*. Paris: L'Harmattan, 1992.

Basagaña, Xavier, et al. "Heat Waves and Cause-Specific Mortality at All Ages." *Epidemiology* 22, no. 6 (2011): 765–72.

Basu, R., and B. Malig. "High Ambient Temperature and Mortality in California: Exploring the Roles of Age, Disease, and Mortality Displacement." *Environmental Research* 111, no. 8 (2011): 1286–92.

Bauer, Alain, and Xavier Raufer. *Violences et insécurité urbaines*. Paris: Presses Universitaires de France, 1998.

Beauvoir, Simone de. *La vieillesse*. Paris: Gallimard, 1970.

Beck, Ulrich. *Risk Society: Towards a New Modernity*. London: Sage, 1992.

———. *World Risk Society*. Cambridge: Polity Press, 1999.

Belmin, Joël. "Les conséquences de la vague de chaleur d'août 2003 sur la mortalité des personnes âgées: Un premier bilan." La presse médicale 32 (18 Oct. 2003): 1591–94.

Belmin, Joël, et al. "Level of Dependency: A Simple Marker Associated with Mortality during the 2003 Heatwave among French Dependent Elderly People Living in the Community or in Institutions." *Age and Ageing* 36 (2007): 298–303.

Benoît-Lapierre, Nicole, Rithée Cevasco, and Markos Zafiropoulos. *Vieillesse des pauvres: Les chemins de l'hospice.* Paris: Éditions Économie et Humanisme, 1980.

Berman, Marshall. *All That Is Solid Melts into Air: The Experience of Modernity.* New York: Penguin, 1982.

Bertillon, Jacques. "De la population de la France et des remèdes à y apporter." *Journal de la société statistique de Paris* 36 (1895): 410–38.

Besancenot, Jean-Pierre. "Vagues de chaleur et mortalité dans les grandes agglomérations urbaines." *Environnement, risques et santé* 1, no. 4 (2002): 229–40.

Bess, Michael. *The Light-Green Society: Ecology and Technological Modernity in France, 1960–2000.* Chicago: University of Chicago Press, 2003.

Bess, Michael, Mark Cioc, and James Sievert. "Environmental History Writing in Southern Europe." *Environmental History* 5, no. 4 (2000): 545–56.

Biehl, João. *Vita: Life in a Zone of Social Abandonment.* Berkeley: University of California Press, 2005.

Blanchot, Amélia. "Pour une chambre de 12 m2, 40 candidats dans l'escalier." *Libération,* 20 September 2007.

Blondet, Maxime. "Les oubliés reposent en terre." *France-Soir,* 4 September 2003.

Bonvalet, Catherine, and Jim Ogg. "The Housing Situation and Residential Strategies of Older People in France." *Ageing & Society* 28 (2008): 753–77.

Boudes, Thierry, and Hervé Laroche. "Taking Off the Heat: Narrative Sensemaking in Post Crisis Inquiry Reports." *Organization Studies* 30 (2009): 377–96.

Bourdelais, Patrice. *L'âge de la vieillesse: Histoire du vieillissement de la population.* Paris: Odile Jacob, 1993.

Bourdieu, Pierre. "Effets de lieu." In *Le misère du monde,* edited by Pierre Bourdieu, 159–67. Paris: Seuil, 1993.

———. "L'illusion biographique." *Actes de la recherche en sciences sociales* 62–63 (1986): 69–72.

Bourgois, Philippe. "Missing the Holocaust: My Father's Account of Auschwitz from August 1943 to June 1944." *Anthropological Quarterly* 78, no. 1 (2005): 89–123.

Boverat, Fernand. *Comment nous vaincrons la dénatalité.* Paris: L'Alliance nationale contre la dépopulation, 1939.

Boyer, L., S. Robitail, D. Debensason, P. Auquier, and J.-L. San Marco. "Média et santé publique: L'exemple de la canicule pendant l'été 2003 en France." *Revue d'épidémiologie et de santé publique* 53 (2005): 525–34.

Braga, Alfésio, Luís Ferreira, Antonella Zanobetti, and Joel Schwartz. "The Time Course of Weather-Related Deaths." *Epidemiology* 12, no. 6 (2001): 662–67.

Brard, Delphine. "La fabrique médiatique de la canicule d'août 2003 comme problème public." DEA mémoire, Université de Paris-I Panthéon-Sorbonne, 2004.

Brubaker, Rogers. *Citizenship and Nationhood in France and Germany.* Cambridge, MA: Harvard University Press, 1992.

Brücker, Gilles. "Impact sanitaire de la vague de chaleur d'août 2003: Premiers résultats et travaux à mener." *Bulletin épidémiologique hebdomadaire* 45–46 (2003): 217.

———. "Le plan national canicule à l'épreuve." *Bulletin épidémiologique hebdomadaire* 22–23 (2007): 189–90.

Buck-Morss, Susan. *The Dialectics of Seeing: Walter Benjamin and the Arcades Project.* Cambridge, MA: MIT Press, 1991.

Buffon, Georges-Louis Leclerc, Comte de. "Des époques de la nature." In *Histoire naturelle, générale et particulière.* Supplement V. Paris, 1778.

Bungener, Martine. "Canicule estivale: La triple vulnérabilité des personnes âgées." *Mouvements* 32 (2004): 75–82.

Butler, Judith. *Frames of War: When Is Life Grievable?* London: Verso, 2010.

———. *Precarious Life: The Powers of Mourning and Violence.* London: Verso, 2004.

Cabirol, Claude. *La condition des personnes âgées: Évolution et aspects actuels.* Toulouse: Privat, 1981.

Cabut, Sandrine. "L'augmentation des décès perceptible dès le 4 août." *Libération*, 26 September 2003.

———. "Autopsie d'une canicule." *Libération*, 9 September 2003.

———. "Bilan d'ici à 15 jours." *Libération*, 20 August 2003.

———. "Tout ce qui avait été mis en place est en train d'être supprimé." *Libération*, 22 August 2003.

Cabut, Sandrine, Eric Favereau, and Julie Lasterade. "Les extraits du rapport et les réponses des professionnels concernés: 'Il ne faut pas chercher des boucs emissaries.'" *Libération*, 9 September 2003.

Cabut, Sandrine, and Nicole Penicaut. "Le nombre de victims pourrait atteindre 10,000." *Libération*, 21 August 2003.

Cadot, Emmanuelle, Victor G. Rodwin, and Alfred Spira. "In the Heat of the Summer: Lessons from the Heat Waves in Paris." *Journal of Urban Health* 84, no. 4: 466–68.

Cadot, Emmanuelle, and Alfred Spira. "Canicule et surmortalité à Paris en août 2003, le poid des facteurs socio-économiques." *Espaces, populations, sociétés* 2–3 (2006): 239–49.

Caldwell, Christopher. "Revolting High Rises: Were the French Riots Produced by Modern Architecture?" *New York Times Magazine*, 27 November 2005.

"Canicule: Un bilan de plus en plus lourd." *Libération*, 30 August 2003.

"Canicule: La liste des 66 morts oubliés." *Le Parisien*, 2 September 2003.

"La canicule de 2006 a fait 133 morts." *Le Nouvel Observateur*, 24 October 2006.

Canouï-Poitrine, F., Emmanuelle Cadot, and Alfred Spira. "Excess Deaths during the August 2003 Heat Wave in Paris, France." *Revue d'épidémiologie et de santé publique* 54 (2006): 127–35.

Caradec, Vincent. "'Seniors' et 'personnes âgées': Réflexions sur les modes de catégorisation de la vieillesse." In *Regards croisés sur la protection sociale de la vieillesse*, edited by Françoise Cribier and Elise Feller, 313–26. Paris: Comité d'histoire de la sécurité sociale, 2005.

———. *Sociologie de la vieillesse et du vieillissement.* Paris: Nathan, 2001.

Castel, Robert. *Les métamorphoses de la question sociale: Une chronique du salariat.* Paris: Fayard, 1995.

Castells, Manuel. *The City and the Grassroots: A Cross-Cultural Theory of Urban Social Movements.* Berkeley: University of California Press, 1983.

Castelnau, Vérane, Mathieu Ecoiffier, Marie-Joëlle Gros, Nicole Pénicaut, and Mehdi Rahaoui. "De l'hôpital au cimetière, la saturation." *Libération*, 14 August 2003.

Chabrun, Laurent, Jérôme Dupuis, Eric Pelletier, and Jean-Marie Pontaut. "Qui étaient les oubliés de la canicule?" *L'Express*, 25 December 2003.

Chaineaud, Christel. "L'État et la veuve âgée sous la IIIe République: Entre indifference et prise en charge." In *De l'hospice au domicile collectif: La vieillesse et ses prises en charge de la fin du XVIIIe siècle à nos jours*, edited by Yannick Marec and Daniel Réguer, 419–27. Mont-Saint-Aignan: Presses Universitaires de Rouen et du Havre, 2013.

Chamak, Raphaël, Alexandre Garcia, and Cécile Prieur. "Ces vieillards morts dans l'oubli à Paris lors de la canicule." *Le Monde*, 4 September 2003.

Champiat, Clément. "Identifier les îlots de chaleur urbains pour réduire l'impact sanitaire des vagues de chaleur." *Environnement, risques et santé* 8, no. 5 (2009): 399–411.

Champigny, André. "Les dix commandements de la vieillesse." *Gérontologie* 8 (1972): 36.

Chauvel, Louis. *Le destin des générations: Structure sociale et cohortes en France au XXe siècle.* Paris: Presses Universitaires de France, 1998.

"Chiffres et confusion." *Libération*, 22 August 2003.

"Le cimetière en état de siege." *Le Parisien.* 4 September 2003.

Clark, T. J. *The Painting of Modern Life: Paris in the Art of Manet and His Followers.* Princeton, NJ: Princeton University Press, 1985.

Clarke, Lee. *Worst Cases: Terror and Catastrophe in the Popular Imagination.* Chicago: University of Chicago Press, 2006.

Clerc, Paul. *Grandes ensembles, banlieues nouvelles: Enquête démographique et psycho-sociologique.* Paris: Centre de Recherche d'Urbanisme, INED, 1967.

Cobb, Richard. *Paris and Elsewhere.* New York: John Murray, 1999.

Coffman, Christine. "The Papin Enigma." *GLQ: A Journal of Lesbian and Gay Studies* 5, no. 3 (1999): 331–59.

Coing, Henri. *Rénovation urbaine et changement social.* Paris: Éditions Ouvrières, 1966.

Cole, Joshua. *The Power of Large Numbers: Population, Politics, and Gender in Nineteenth-Century France.* Ithaca, NY: Cornell University Press, 2000.

Coleman, William. *Death Is a Social Disease: Yellow Fever in the North.* Madison: University of Wisconsin Press, 1982.

Collier, Stephen J. "Topologies of Power: Foucault's Analysis of Political Government beyond Governmentality." *Theory, Culture, and Society* 26 (2009): 78–108.

Comfort, L., et al. "Reframing Disaster Policy: The Global Evolution of Vulnerable Communities." *Environmental Hazards* 1 (1999): 39–44.

"Le comique de Bercy." *Le Canard Enchaîné*, 6 August 2003.

Cooper, Melinda. *Life as Surplus: Biotechnology and Capitalism in the Neoliberal Era.* Seattle: University of Washington Press, 2008.

Corbin, Alain. *Le miasme et la jonquille.* Paris: Flammarion, 1982.

"Courrier." *La Croix*, 13 October 2003.

Cribier, Françoise. "Vieillesse et citoyenneté." In *Villes et vieillir*, 312–20. Paris: La Documentation française, 2004.

Cronon, William. "Introduction: In Search of Nature." In *Uncommon Ground: Toward Reinventing Nature*, edited by William Cronon, 23–56. New York: W.W. Norton, 1995.

Cupers, Kenny. "Designing Social Life: The Urbanism of the *Grands Ensembles.*" *Positions* 1 (2010): 94–121.

———. "The Expertise of Participation: Mass Housing and Urban Planning in Post-war France." *Planning Perspectives* 26, no. 1 (2011): 29–53.

Cusack, Lynette, Charlotte de Crespigny, and Peter Athanasos. "Heatwaves and Their Impact on People with Alcohol, Drug, and Mental Health Conditions: A Discussion Paper on Clinical Practice Considerations." *Journal of Advanced Nursing* 67, no. 4 (2011): 915–22.

Cutter, Susan L. *American Hazardscapes: The Regionalization of Hazards and Disasters.* Washington, DC: Joseph Henry Press, 2001.

———. "The Vulnerability of Science and the Science of Vulnerability." *Annals of the Association of American Geographers* 93, no. 1 (2003): 1–12.

Cutter, Susan L., Bryan J. Boruff, and W. Lynn Shirley. "Social Vulnerability to Environmental Hazards." *Social Science Quarterly* 84, no. 2 (2003): 242–61.

Damon, Julien. "La citoyenneté du SDF." DEA thesis, Université de Paris IV, 1994.

Daniels, Ronald J., Donald F. Kettl, and Howard Kunreuther. *On Risk and Disaster: Lessons from Hurricane Katrina.* Philadelphia: University of Pennsylvania Press, 2006.

Daric, Jean. *Vieillissement de la population et prolongation de la vie active.* Travaux et Documents, Institut National d'Études Démographiques, Cahier no. 7. Paris: Presses Universitaires de France, 1948.

Davis, Mike. *Ecology of Fear: Los Angeles and the Imagination of Disaster.* New York: Vintage, 1998.

———. *Late Victorian Holocausts: The El Niño Famines and the Making of the Third World.* London: Verso, 2002.

de Grazia, Victoria. *Irresistible Empire: America's Advance through 20th-Century Europe.* Cambridge, MA: Harvard University Press, 2005.

Delpiroux, Dominique. "La canicule tue, la polémique s'installe." *La Dépêche*, 12 August 2003.

"Deux mois de chaleur." *Libération*, 9 September 2003.

Di Chiro, Giovanna. "Nature as Community: The Convergence of Environment and Social Justice." In *Uncommon Ground: Toward Reinventing Nature*, edited by William Cronon, 298–320. New York: W. W. Norton, 1995.

Dikeç, Mustafa. "Revolting Geographies: Urban Unrest in France." *Geography Compass* 1, no. 5 (2007): 1190–1206.

Dumons, Bruno, and Gilles Pollet. "La question des retraites vue par les socialistes français (1880–1956)." In *Regards croisés sur la protection sociale de la vieillesse*, edited by Françoise Cribier and Elise Feller, 197–224. Paris: Comité d'histoire de la sécurité sociale, 2005.

Duneier, Mitchell. "Ethnography, the Ecological Fallacy, and the 1995 Chicago Heat Wave." *American Sociological Review* 71 (2006): 679–88.

Dupont-Monod, Clara. "Le secret de Marie France." *Marianne* 334 (15–21 September 2003): 22–23.

Dupuy, Gérard. "Négligence." *Libération*, 3 September 2003.

———. "Sieste." *Libération*, 12 August 2003.

Dupuy, Jean-Pierre. *Petite métaphysique des tsunamis.* Paris: Seuil, 2005.

Dutton, Paul V. *Origins of the French Welfare State: The Struggle for Social Reform in France, 1914–1947.* New York: Cambridge University Press, 2002.

Dyson, Michael Eric. *Come Hell or High Water: Hurricane Katrina and the Color of Disaster.* New York: Basic, 2006.

Ebbeson, Jonas, ed. *Access to Justice in Environmental Matters in the EU.* The Hague: Kluwer Law International, 2002.

Ecoiffier, Matthieu. "Canicule: Mattei se defend sans gloire." *Libération*, 12 September 2003.

———. "William Dab, promotion 'cocasse.'" *Libération*, 22 August, 2003.

"Eden Bio." *Smart Urbanism.* http://www.smarturb.org/index.php?#article:93.

"Eden Bio by Édouard François." *De Zeen.* http://www.dezeen.com/2009/02/17/eden-bio-by-edouard-francois-2/.

Epstein, Steven. *Impure Science: AIDS, Activism, and the Politics of Knowledge.* Berkeley: University of California Press, 1996.

Equy, Laure. "Sous la douche, je pense à l'endroit où je vais dormir." *Libération*, 22 January 2008.

Erikson, Kai. *A New Species of Trouble: Explorations in Disaster, Trauma, and Community.* New York: Norton, 1994.

Erlanger, Steven. "Across France, Café Owners Are Suffering." *New York Times*, 22 November 2008.

Esposito, Roberto. *Bíos: Biopolitics and Philosophy*. Translated by Timothy Campbell. Minneapolis: University of Minnesota Press, 2008.

Estrosi, France. "L'intéraction avant tout." In *Le sentiment d'insécurité: Les seniors face à la délinquance*. Paris: Éditions Taitbout, 1994.

"L'État finit par entendre le SOS des médecins." *Libération*, 14 August 2003.

Evenson, Norma. *Paris: A Century of Change, 1878–1978*. New Haven, CT: Yale University Press, 1979.

Ewald, François. *L'État-providence*. Paris: Grasset, 1986.

Farge, Arlette. *Fragile Lives: Violence, Power, and Solidarity in Eighteenth-Century Paris*. Translated by Carole Shelton. Cambridge, MA: Harvard University Press, 1993.

Farmer, Paul. "On Suffering and Structural Violence: A View from Below." *Daedalus* 125 (1996): 261–83.

Fassin, Didier. *La force de l'ordre: Une anthropologie de la police des quartiers*. Paris: Seuil, 2011.

Fassin, Didier, and Anne-Jeanne Naudé. "Plumbism Reinvented: Childhood Lead Poisoning in France, 1985–1990." *American Journal of Public Health* 94 (2004): 1854–63.

Fassin, Eric, and Didier Fassin. *Question sociale, question raciale?* Paris: Éditions la Découverte, 2006.

Favereau, Eric. "Le 11 août tel que l'ont vécu trois acteurs clés de la santé." *Libération*, 25 September 2003.

———. "La canicule menace aussi les malades mentaux." *Libération*, 4 September 2003.

———. "Je saurai comment faire." *Libération*, 12 September 2003.

———. "Mattei, ministre en mauvais santé." *Libération*, 21 August 2003.

———. "Pourquoi Mattei na rien vu venir?" *Libération*, 25 September 2003.

Feller, Elise. "L'entrée en politique d'un groupe d'âge: La lutte des pensionnés de l'État dans l'entre-deux-guerres et la construction d'un 'modèle français' de retraite." *Le mouvement social* 190 (2000): 33–59.

———. "Les femmes et le vieillissement dans la France du premier XXe siècle." *Clio* 7 (1998): 199–222.

———. *Histoire de la vieillesse en France, 1900–1960: Du vieillard au retraité*. Paris: Seli Arslan, 2005.

Fijalkow, Yankel. "Surpopulation ou insalubrité: Deux statistiques pour décrire l'habitat populaire (1880–1914)." *Le mouvement social* 182 (1998): 79–96.

Fineman, Joel. "The History of the Anecdote: Fiction and Fiction." In *The New Historicism*, edited by Harold Aram Veeser, 49–76. New York: Routledge, 1989.

Forbes, Jill. "Bourdieu's Maieutics." *SubStance* 29, no. 3 (2000): 22–42.

Ford, Caroline. "Landscape and Environment in French Historical and Geographical Thought: New Directions." *French Historical Studies* 24 (2001): 125–34.

———. "Nature, Culture, and Conservation in France and Her Colonies, 1840–1940." *Past and Present* 183 (2004): 173–98.

Fortun, Kim. *Advocacy after Bhopal: Environmentalism, Disaster, New Global Orders*. Chicago: University of Chicago Press, 2001.

Foucault, Michel. *Discipline and Punish: A History of the Prison*. Translated by Alan Sheridan. New York: Vintage, 1979.

———. "Governmentality." Translated by Rosi Braidotti and revised by Colin Gordon. In *The*

Foucault Effect: Studies in Governmentality, edited by Graham Burchell, Colin Gordon, and Peter Miller,87–104. Chicago: University of Chicago Press.
———. *The History of Sexuality.* Volume 1, *An Introduction.* New York: Vintage, 1981.
———. *Security, Territory, Population: Lectures at the Collège de France, 1977–1978.* Translated by Graham Burchell. New York: Picador, 2007.
———. *"Society Must Be Defended": Lectures at the Collège de France, 1975–1976.* Translated by David Macey. New York: Picador, 2003.
———. "La vie des hommes infâmes." In *Dits et écrits, 1954–1988*, vol. 3, edited by Daniel Defert and François Ewald, 237–53. Paris: Gallimard, 1994.
Foucras, Philippe, and Christian Lehmann. "Salauds de toubibs!" *Libération*, 11 September 2003.
Fouillet, Anne. "Comparaison de la surmortalité observée en juillet 2006 à celle estimée à partir des étés 1975–2003, France." *Bulletin épidémiologique hebdomadaire* 22–23 (2007): 192–97.
Franc, Robert. *Le scandale de Paris.* Paris: Grasset, 1971.
"Francis Mer, devin d'honneur." *Le Canard Enchaîné*, 20 August 2003.
Fredman, Stephen. "'How to Get Out of the Room That Is the Book?' Paul Auster and the Consequences of Confinement." *Postmodern Culture* 6, no. 3 (1996). muse.jhu.edu/journals/postmodern_culture/v006/6.3fredman.html.
Furlough, Ellen. "Making Mass Vacations: Tourism and Consumer Culture in France, 1930s-1970s." *Comparative Studies in Society and History* 40, no. 2 (1998): 247–86.
Gaillard, Jeanne. *Paris, la ville (1852–1870).* Paris: L'Harmattan, 1977.
Gallagher, Catherine, and Stephen Greenblatt. *Practicing New Historicism.* Chicago: University of Chicago Press, 2001.
Gandy, Matthew. "Cyborg Urbanization: Complexity and Monstrosity in the Contemporary City." *International Journal of Urban and Regional Research* 29, no. 1 (2005): 26–49.
———. "The Ecological Facades of Patrick Blanc." *Architectural Design* 80, no. 3 (2010): 28–33.
———. "The Paris Sewers and the Rationalization of Urban Space." *Transactions of the Institute of British Geographers*, n.s. 24 (1999): 23–44.
———. "Zones of Indistinction: Bio-Political Contestations in the Urban Arena." *Cultural Geographies* 13 (2006): 497–516.
Gavalda, Anna. *Ensemble, c'est tout.* Paris: Le dilettante, 2004.
Geertz, Clifford. *The Interpretation of Cultures.* New York: Basic Books, 1973.
Gelfand, Toby. "La canicule: Paris 2003." *Analecta historico medica: Memorias 41 Congreso internacional de historia de la medicina*, Suplemento 1 (2008): 175–80.
Giddens, Anthony. *The Consequences of Modernity.* Stanford, CA: Stanford University Press, 1990.
Greenblatt, Stephen. "The Touch of the Real." *Representations* 59 (1997): 14–29.
Grémy, I., et al. "Conséquences sanitaires de la canicule d'août 2003 en Île-de-France: Premier bilan." *Revue d'épidémiologie et de santé publique* 52 (2004): 93–108.
Gros, Marie-Joélle, and Nicole Penicaut. "Viellesse: La grande hypocrise de Raffarin." *Libération*, 18 August 2003.
Guillemard, Anne-Marie. *Le déclin du social: Formation et crise des politiques de la vieillesse.* Paris: Presses Universitaires de France, 1986.
Guillemin, Jeanne. *Anthrax: The Investigation of a Deadly Outbreak.* Berkeley: University of California Press, 1999.
Hajat, S., et al. "Mortality Displacement of Heat-Related Deaths: A Comparison of Delhi, São Paulo, and London." *Epidemiology* 16, no. 5 (2005): 613–20.

Hardt, Michael, and Antonio Negri. *Empire*. Cambridge, MA: Harvard University Press, 2000.

Harvey, David. *The Condition of Postmodernity*. Oxford: Blackwell, 1990.

———. *Consciousness and the Urban Experience: Studies in the History and Theory of Capitalist Urbanization*. Baltimore: Johns Hopkins University Press, 1985.

———. *Social Justice and the City*. Oxford: Blackwell, 1988.

Hémon, Denis, et al. "Surmortalité liée à la canicule d'août 2003 en France." *Bulletin épidémiologique hebdomadaire* 45–46 (2003): 221–25.

Henry, Louis, and Maurice Febvay. "La situation du logement dans la région parisienne." *Population* 12, no. 1 (1957): 129–40.

Henry, Michel. "Trop d'indifférence face à la mortalité des vieux." *Libération*, 12 August 2003.

Henson, Bob. "The Weather Notebook." Interview with Eric Klinenberg, 8 July 2004. http://www.weathernotebook.org/transcripts/2004/07/08.php.

Holstein, Josiane, et al. "Were Less Disabled Patients the Most Affected by 2003 Heat Wave in Nursing Homes in Paris, France?" *Journal of Public Health* 27, no. 4 (2005): 359–65.

Houellebecq, Michel. *La possibilité d'une île*. Paris: Fayard, 2005.

Huang, C., et al. "Projecting Future Heat-Related Mortality under Climate Change Scenarios: A Systematic Review." *Environmental Health Perspectives* 119, no. 12 (2011): 681–90.

Hunt, Nancy Rose. "An Acoustic Register, Tenacious Images, and Congolese Scenes of Rape and Repetition." *Cultural Anthropology* 23, no. 2 (2008): 220–53.

Hurley, Andrew. "Creating Ecological Wastelands: Oil Pollution in New York City, 1870–1900." *Journal of Urban History* 20, no. 3 (1994): 340–64.

Huynen, Maud M. T. E., et al. "The Impact of Heat Waves and Cold Spells on Mortality Rates in the Dutch Population." *Environmental Health Perspectives* 109, no. 5 (2001): 463–70.

Israelewicz, Étienne. "Effet moisson: L'impact des catastrophes vie sur la mortalité à long terme." *Bulletin français d'actuariat* 12, no. 24 (2012): 113–59.

"Italy Puts 2003 Heat Toll at 20,000." *International Herald Tribune*, 28 June 2005.

Jackson, Jeffrey H. *Paris under Water: How the City of Light Survived the Great Flood of 1910*. London: Palgrave Macmillan, 2010.

Jackson, Julian. *The Popular Front in France: Defending Democracy, 1934–38*. New York: Cambridge University Press, 1988.

Jacobs, Jane. *The Death and Life of Great American Cities*. New York: Vintage, 1961.

Jacobs, Jane M., Stephen Cairns, and Ignaz Strebel. "'A Tall Storey . . . but, a Fact Just the Same': The Red Road High-Rise as a Black Box." *Urban Studies* 44, no. 3 (2007): 609–29.

Jobs, Richard I. *Riding the New Wave: Youth and the Rejuvenation of France after the Second World War*. Stanford, CA: Stanford University Press, 2007.

Jonquet, Thierry. *Mon vieux*. Paris: Seuil, 2004.

Jordan, David. "Haussmann and Haussmannization: The Legacy for Paris." *French Historical Studies* 27, no. 1 (2004): 87–113.

———. *Transforming Paris: The Life and Labors of Baron Haussmann*. New York: Free Press, 1995.

Jorland, Gérard, Annick Opinel, and George Weisz, eds. *Body Counts: Medical Quantification in Historical and Sociological Perspective/La quantification médicale, perspectives historiques et sociologiques*. Montreal: McGill/Queen's University Press, 2005.

Kain, Alexandra. "Eden Bio: Paris Grows a Green Heart." *Inhabitat*, 5 May 2009. http://inhabitat.com/eden-bio.

Klassen, Sherri. "Greying in the Cloister: The Ursuline Life Course in Eighteenth-Century France." *Journal of Women's History* 12, no. 4 (2001): 87–112.

Klinenberg, Eric. "Blaming the Victims: Hearsay, Labeling, and the Hazards of Quick-Hit Disaster Ethnography." *American Sociological Review* 71 (2006): 689–98.

———. "Dying Alone: The Social Production of Urban Isolation. *Ethnography* 2, (2001): 501–31.

———. *Heat Wave: A Social Autopsy of Disaster in Chicago.* Chicago: University of Chicago Press, 2002.

———. "The Politics of Heat Waves: Victims of a Hot Climate and a Cold Society." *International Herald Tribune*, 22 August 2003.

Kniebiehler, Yvonne. *La révolution maternelle: Femmes, modernité, citoyenneté depuis 1945.* Paris: Perrin, 1997.

Koos, Cheryl. "Gender, Anti-Individualism, and Nationalism: The Alliance Nationale and the Pronatalist Backlash against the *Femme moderne*, 1933–1940." *French Historical Studies* 19, no. 3 (1996): 699–723.

Koppe, Christina, Sari Kovats, Gerd Jendritzky, and Bettina Menne. *Heat-Waves: Risks and Responses.* Copenhagen: World Health Organization, 2004.

Kovats, R. S., and S. Hajat. "Heat Stress and Public Health: A Critical Review." *Annual Review of Public Health* 29 (2008): 41–55.

Kuisel, Richard. *Seducing the French: The Dilemma of Americanization.* Berkeley: University of California Press, 1993.

Lacan, Jean-François. *Scandales dans les maisons de retraite.* Paris: Albin Michel, 2002.

Lagadec, Patrick, and Hervé Laroche. *Retour sur les rapports d'enquête et d'expertise suite à la canicule de l'été 2003.* Grenoble: Maison des Sciences de l'Homme-Alpes, 2005.

Landal, Eric. "Sarkozy, pompier chez les pompiers." *Libération*, 29 September 2003.

Laske, Karl. "Un enfant sans papiers fuit la police et chute du 4e étage." *Libération*, 10 August 2007.

———. "Le saut dans le désespoir des sans papiers traqués." *Libération*, 21 September 2007.

Lasterade, Julie. "Noyades accidentelles: + 39 % pendant l'été." *Libération*, 23 September 2003.

Latour, Bruno. *The Pasteurization of France.* Translated by Alan Sheridan and John Law. Cambridge, MA: Harvard University Press, 1988.

Laurent, Éloi. "Bleu, Blanc . . . Green? France and Climate Change." *French Politics, Culture & Society* 27, no. 2 (2009): 142–53.

Lavedan, Pierre. *Nouvelle histoire de Paris: Histoire de l'urbanisme à Paris.* Paris: Hachette, 1975.

Ledrans, Martine, et al. "La vague de chaleur d'août 2003: Que s'est-il passé?" *La revue du praticien* 54 (2004): 1289–97.

Le Grand-Sébille, Catherine, and Anne Véga. *Pour une autre mémoire de la canicule: Professionnels de funéraire, des chambres mortuaires et familles témoignent.* Paris: Vuibert, 2005.

Lenoir, Rémi. "L'invention du 'troisième âge'." *Actes de la recherche en sciences sociales* 26–27 (1979): 57–82.

Lentschner, Keren. "Avis de pénurie sur le marché des chambres de bonne." *Le Figaro*, 9 April 2007.

Le Tertre, Alain, et al. "Première estimation de l'impact de la vague de chaleur sur la mortalité durant l'été 2006." *Bulletin épidémiologique hebdomadaire* 22–23 (2007): 190–92.

Li, Bo, et al. "The Impact of Extreme Heat on Morbidity in Milwaukee, Wisconsin." *Climatic Change* 110, nos. 3–4 (2011): 959–76.

Lindon, Mathieu. "On a oublié d'arroser les vieux!" *Libération*, 30 August 2003.

Löfgren, Orvar. *On Holiday: A History of Vacationing.* Berkeley: University of California Press, 1999.

Mallon, Isabelle. *Vivre en maison de retraite: Le dernier chez-soi.* Rennes: Presses Universitaires de Rennes, 2004.

Marcelle, Pierre. "D'un été meurtrier." *Libération,* 25 August 2003.

Marec, Yannick, and Daniel Réguer, eds. *De l'hospice au domicile collectif: La vieillesse et ses prises en charge de la fin du XVIIIe siècle à nos jours.* Mont-Saint-Aignan: Presses Universitaires de Rouen et du Havre, 2013.

Marshall, T. H. "Citizenship and Social Class." In *Inequality and Society,* edited by Jeff Manza and Michael Sauder, 148–54. New York: Norton, 2009.

Marty, Simon. "Philippe Heurteaux, le clochard millionaire." *Marianne* 334 (15–21 September 2003): 26–27.

Marx, Karl. *Capital: A Critique of Political Economy.* Translated by Ben Fowkes. 3 vols. New York: Penguin, 1976.

"Mattei ni responsible ni coupable." *Libération,* 15 August 2003.

Mauco, Georges. "L'apport humain et social des personnes âgées." *La Liberté,* 13 May 1963.

Maurus, Véronique. "La bataille du 18e." *Le Monde,* 12 September 2003.

McNeill, Donald. "Skyscraper Geography." *Progress in Human Geography* 29, no. 1 (2005): 41–55.

McNeill, J. R. "Observations on the Nature and Culture of Environmental History." *History and Theory* 42 (2003): 5–43.

Meehl, Gerald A., and Claudia Tebaldi. "More Intense, More Frequent, and Longer Lasting Heat Waves in the 21st Century." *Science* 305 (2004): 994–97.

Melosi, Martin V. "Cities, Technological Systems, and the Environment." *Environmental History Review* 14, nos. 1–2 (1990): 45–64.

———. *Garbage in the Cities: Refuse, Reform, and the Environment.* College Station: Texas A&M Press, 1981.

———. "The Place of the City in Environmental History." *Environmental History Review* 17, no. 1 (1993): 1–24.

Mercier, Louis-Sébastien. *Tableau de Paris.* Hamburg: Virchaux; Neuchâtel: S. Fauche, 1781.

Meyer, William B. "Urban Heat Island and Urban Health: Early American Perspectives." *Professional Geographer* 43 (1991): 38–48.

Michaud Nérard, François. *La révolution dans la mort.* Paris: Vuibert, 2007.

Milet, Marc. "Cadres de perception et luttes d'imputation dans la gestion de crise: L'exemple de 'la canicule' d'août 2003." *Revue français de science politique* 55, no. 4 (2005): 573–605.

Mileti, Dennis. *Disasters by Design: A Reassessment of Natural Hazards in the United States.* Washington, DC: Joseph Henry Press, 1999.

Milgram, Stanley. "The Experience of Living in Cities." *Science* 167, no. 3924 (1970): 1461–68.

Mitman, Gregg. "In Search of Health: Landscape and Disease in American Environmental History." *Environmental History* 10 (2005): 184–210.

Morali, Delphine, Louis Jehel, and Sabrina Paterniti. "Étude de l'influence de la canicule d'août 2003 sur la fréquence des consultations de psychiatrie d'urgence et les comportements suicidaires." *La presse médicale* 37, no. 2 (2008): 224–28.

Morizet, André. *Du vieux Paris au Paris moderne.* Paris: Hachette, 1932.

"Morts: Mattei dévoile ses chiffres." *Libération,* 29 August 2003.

Newsome, W. Brian. *French Urban Planning, 1940–1968: The Construction and Deconstruction of an Authoritarian System.* New York: Peter Lang, 2009.

"Nous devrions avoir honte." *Le Parisien,* 26 August 2003.

Nord, Philip. *France's New Deal: From the Thirties to the Postwar Era.* Princeton, NJ: Princeton University Press, 2010.

Nye, Robert. *Crime, Madness, and Politics in Modern France: The Medical Concept of National Decline.* Princeton, NJ: Princeton University Press, 1984.

Ojakangas, Mika. "Impossible Dialogue on Bio-Power: Agamben and Foucault." *Foucault Studies* 2 (2005): 5–28.

O'Rourke, Morgan. "The Property/Casualty Market in 2011." *Risk Management*, February 2011.

Parant, Alain. "Croissance démographique et vieillissement." *Population* 47, no. 6 (1992): 1657–76.

Payet, Marc. "Le dernier hommage aux oubliés de la canicule." *Le Parisien*, 4 September 2003.

———. "Je tenais à être là." *Le Parisien*, 4 September 2003.

Payet, Marc, and Charles de Saint-Sauveur. "Canicule: L'histoire." *Le Parisien*, 14 August 2003.

Pedersen, Susan. *Family, Dependence, and the Origins of the Welfare State: Britain and France, 1914–45.* New York: Cambridge University Press, 1993.

Pelloux, Patrick. *Urgentiste.* Paris: Fayard, 2004.

Pereira, Acacio, and Patrick Roger. "Canicule: En région parisienne, quelque 300 corps n'ont toujours pas èté réclamés par leur famille." *Le Monde*, 26 August 2003.

Perez, Martine. "Canicule: 14,802 décès en France." *Le Figaro*, 26 September 2003.

———. "Canicule 2003: 70,000 morts en Europe." *Le Figaro*, 15 October 2007.

———. "Le mois d'août meurtrier avait fait 20.000 de morts en Italie; La vérité révélée avec deux ans de retard." *Le Figaro*, 28 June 2005.

Perrow, Charles. *Normal Accidents: Living with High-Risk Technologies.* Princeton, NJ: Princeton University Press, 1999.

"Les personnes âgées ont-elles oui ou non le droit de vivre?" *France-Soir*, 20 June 1962.

Peters, Terri. "Housing on Rue des Vignoles." *Architecture Week*, 17 March 2010. http://www.ArchitectureWeek.com/2010/0317/index.html.

"Le petit polo de Mattei." *Le Parisien*, 31 December 2003.

Petryna, Adriana. *Life Exposed: Biological Citizens after Chernobyl.* Princeton, NJ: Princeton University Press, 2002.

Pick, Daniel. *Faces of Degeneration: A European Disorder.* New York: Cambridge University Press, 1993.

Picon-Lefebvre, Virginie. *Paris-ville moderne: Maine-Montparnasse et la Défense, 1950–1975.* Paris: Norma, 2003.

Pinkney, David. *Napoleon III and the Rebuilding of Paris.* Princeton, NJ: Princeton University Press, 1958.

Porquery, Didier. "Spéculations." *Libération*, 20 September 2007.

Posner, Richard A. *Catastrophe: Risk and Response.* New York: Oxford University Press, 2004.

Pouchelle, Marie-Christine. *Vivre dans un grand ensemble: "Les cimentiers."* Paris: Epi, 1974.

Pritchard, Sara B. "Reconstructing the Rhône: The Cultural Politics of Nature and Nation in Contemporary France, 1945–1997." *French Historical Studies* 27 (2004): 765–99.

Proctor, Robert N., and Londa Schiebinger, eds. *Agnotology: The Making and Unmaking of Ignorance.* Stanford, CA: Stanford University Press, 2008.

"Quand l'égoïsme et l'indifférence ont tué." *Le Figaro*, 28 August 2003.

Quinio, Dominique. "Aux morts inconnus." *La Croix*, 3 September 2003.

Rabinow, Paul. *French DNA.* Chicago: University of Chicago Press, 1997.

Redfield, Peter. *Life in Crisis: The Ethical Journey of Doctors without Borders.* Berkeley: University of California Press, 2012.

"René Hamaoui, inconnu à son adresse." *Marianne* 334 (15–21 September 2003): 25.

Rey, Grégoire. "Vagues de chaleur, fluctuations ordinaires des températures et mortalité en France depuis 1971." *Population* 62, no. 3 (2007): 533–63.

Rey, Grégoire, et al. "Heat Exposure and Socio-Economic Vulnerability as Synergistic Factors in Heat-Wave-Related Mortality." *European Journal of Epidemiology* 24 (2009): 495–502.

Robcis, Camille. *The Law of Kinship: Anthropology, Psychoanalysis, and the Family in France.* Ithaca, NY: Cornell University Press, 2013.

Robert-Diard, Pascale. "Le cadavre, la cour d'appel et le trouble anormal de voisinage." *Le Monde,* 18 May 2009.

Roberts, Mary Louise. *Civilization without Sexes: Reconstructing Gender in Postwar France, 1914–1939.* Chicago: University of Chicago Press, 1994.

Robine, Jean-Marie, et al. "Death Toll Exceeded 70,000 in Europe during the Summer of 2003." *Comptes rendus: Biologies* 331 (2008): 171–78.

Roché, Sebastian. *Sociologie politique de l'insécurité: Violences urbaines, inégalités et globalization.* Paris: Presses Universitaires de France, 1998.

Rose, Nikolas. *The Politics of Life Itself: Biomedicine, Power, and Subjectivity in the Twenty-First Century.* Princeton, NJ: Princeton University Press, 2007.

Rose, Nikolas, Pat O'Malley, and Mariana Valverde. "Governmentality." *Annual Review of Law and Social Science* 2 (2006): 83–104.

Rosen, Christine Meisner, and Joel Arthur Tarr. "The Importance of an Urban Perspective in Environmental History." *Journal of Urban History* 20, no. 3 (1994): 299–310.

Rosenberg, Charles. "Cholera in Nineteenth-Century Europe: A Tool for Social and Economic Analysis." *Comparative Studies in Society and History* 8 (1966): 452–63.

———. "Pathologies of Progress: The Idea of Civilization as Risk." *Bulletin of the History of Medicine* 72 (1998): 714–30.

Ross, Kristin. *Fast Cars, Clean Bodies: Decolonization and the Reordering of French Culture.* Cambridge, MA: MIT Press, 1996.

———. "Schoolteachers, Maids, and Other Paranoid Histories." *Yale French Studies* 91 (1997): 7–27.

Rousseau, Daniel. "Surmortalité des étés caniculaires et surmortalité hivernale en France." *Climatologie* 3 (2006): 43–54.

Royer, Solenn de. "Les 'oubliés de la canicule' avaient presque tous de la famille." *La Croix,* 20 October 2003.

Rusnock, Andrea. *Vital Accounts: Quantifying Health and Population in Eighteenth-Century England and France.* New York: Cambridge University Press, 2002.

Saint-Sauveur, Charles de. "Canicule: Les victimes de la chaleur de plus en plus nombreuses." *Le Parisien,* 10 August 2003.

"Sanitaire sans precedent." *Le Parisien,* 15 August 2003.

Schär, Christoph, et al. "The Role of Increasing Temperature Variability in European Summer Heatwaves." *Nature* 427 (2004): 332–36.

Scheper-Hughes, Nancy. *Death without Weeping: The Violence of Everyday Life in Brazil.* Berkeley: University of California Press, 1992.

Schnall, David, and Jean Brami. "Décès à domicile dus à la canicule: Enquête sure les décès survenus en août 2003 dans le 19e arrondissement de Paris." *La revue du practicien—Médecine générale* 18 (2004): 1007–11.

Schneider, Vanessa. "Le gouvernement poursuit la chasse aux fainéants." *Libération,* 29 August 2003.

Schneider, Vanessa, and François Wenz-Dumas. "Les sept erreurs capitals du gouvernement." *Libération*, 30 August 2003.

Schwartz, Timothy T. *Building Assessments and Rubble Removal in Quake-Affected Neighborhoods in Haiti: BARR Survey Final Report.* Washington, DC: USAID, 2011.

Scott, Joan Wallach. "French Universalism in the Nineties." *Differences* 15, no. 2 (2004): 32–53.

———. *Parité! Sexual Equality and the Crisis of French Universalism.* Chicago: University of Chicago Press, 2005.

Segalen, Martine. "Les changements familiaux depuis le début du XXe siècle." In *Histoire de la population française*, edited by Jacques Dupâquier et al., 4: 499–541. Paris: Presses Universitaires de France, 1988.

Semenza, Jan C., et al. "Heat-Related Deaths during the July 1995 Heat Wave in Chicago." *New England Journal of Medicine* 335 (1996): 84–90.

Sen, Amartya. *Poverty and Famines: An Essay on Entitlement and Deprivation.* Oxford: Oxford University Press, 1981.

Serafini, Tonino. "Encore à la rue." *Libération*, 1 November 2007.

———. "Les mal-logés investissent la rue de la Banque." *Libération*, 5 October 2007.

———. "Une ville étranglée par la crise du logement." *Libération*, 11 September 2007.

Shapiro, Ann-Louise. *Housing the Poor of Paris: 1850–1902.* Madison: University of Wisconsin Press, 1985.

Shea, Katherine M. "Global Climate Change and Children's Health." *Pediatrics* 120 (2007): 1359–67.

Sirinelli, Jean-François. "Appel des intellectuels en soutien aux grévistes." *Libération*, 12 January 1998.

———. *Les baby-boomers: Une génération, 1945–1969.* Paris: Fayard, 2003.

Smith, Timothy B. *France in Crisis: Welfare, Inequality and Globalization since 1980.* New York: Cambridge University Press, 2004.

Smoyer, Karen E. "Putting Risk in Its Place: Methodological Considerations for Investigating Extreme Event Health Risk." *Social Science and Medicine* 47, no. 11 (1998): 1809–24.

Soysal, Yasemin. *The Limits of Citizenship: Migrants and Postnational Membership in Europe.* Chicago: University of Chicago Press, 1994.

Spary, Emma C. *Utopia's Garden: French Natural History from the Old Regime to Revolution.* Chicago: University of Chicago Press, 2000.

Stearns, Peter. *Old Age in European Society: The Case of France.* London: Croom Helm, 1977.

Steinberg, Ted. *Acts of God: The Unnatural History of Natural Disaster in America.* New York: Oxford University Press, 2006.

———. "The Secret History of Natural Disaster." *Environmental Hazards* 3 (2001): 31–34.

Stine, Jeffrey K., and Joel A. Tarr. "At the Intersection of Histories: Technology and the Environment." *Technology and Culture* 39, no. 4 (1998): 601–40.

Stott, Peter A., et al. "Human Contribution to the European Heat Wave of 2003." *Nature* 432 (2004): 610–14.

Stovall, Tyler. "From Red Belt to Black Belt: Race, Class, and Urban Marginality in Twentieth-Century Paris." In *The Color of Liberty: Histories of Race in France*, edited by Sue Peabody and Tyler Stovall, 351–70. Durham, NC: Duke University Press, 2003.

Sunder Rajan, Kaushik. *Biocapital: The Constitution of Postgenomic Life.* Durham, NC: Duke University Press, 2006.

———. *The Rise of the Paris Red Belt.* Berkeley: University of California Press, 1990.

Sutcliffe, Anthony. *The Autumn of Central Paris: The Defeat of Town Planning, 1850–1970*. London: Edward Arnold, 1970.

Tarr, Joel A., and Gabriel Dupuy. *Technology and the Rise of the Networked City in Europe and America*. Philadelphia: Temple University Press, 1988.

Thenard, Jean-Michel. "Glaciation." *Libération*, 2 September 2003.

Thoraval, Armelle. "Canicule: Les chiffres imprecise des pompiers." *Libération*, 20 September 2003.

Toulemon, Laurent, and Magali Barbieri. "The Mortality Impact of the August 2003 Heat Wave in France: Investigating the 'Harvesting Effect' and Other Long-Term Consequences." *Population Studies* 62, no. 1 (2008): 39–53.

Treichler, Paula. *How to Have Theory in an Epidemic: Cultural Chronicles of AIDS*. Durham, NC: Duke University Press, 1999.

Trincaz, Jacqueline. "Les fondements imaginaires de la vieillesse dans la pensée occidentale." *L'homme* 38 (1999): 167–89.

Troyansky, David. *Old Age in the Old Regime: Image and Experience in Eighteenth-Century France*. Ithaca, NY: Cornell University Press, 1989.

Turner, Bryan S. "Outline of a Theory of Citizenship." In *Citizenship: Critical Concepts*, edited by Bryan S. Turner and Peter Hamilton, 199–226. New York: Routledge, 1994.

Valleron, Alain-Jacques, and Ariane Boumendil. "Épidémiologie et canicules: Analyses de la vague de chaleur 2003 en France." *Comptes rendus: Biologies* 327 (2004): 1125–41.

Vandentorren, Stephanie, et al. "Données météorologiques et enquêtes sur la mortalité dans 13 grandes villes françaises." *Bulletin épidémiologique hebdomadaire* 45–46 (2003): 219–20.

———. "Mortality in 13 French Cities during the August 2003 Heat Wave." *American Journal of Public Health* 94, no. 9 (2004): 1518–20.

Van Loon, Joost. *Risk and Technological Culture: Towards a Sociology of Virulence*. New York: Routledge, 2002.

Vassy, Carine, Richard Keller, and Robert Dingwall. *Enregistrer les morts, identifier les surmortalités: Une comparaison Angleterre, États-Unis et France*. Rennes: Presses de l'EHESP, 2010.

Vaughan, Diane. *The Challenger Launch Decision: Risky Technology, Culture, and Deviance at NASA*. Chicago: University of Chicago Press, 1997.

Vayssière, Bruno-Henri. *Reconstruction, deconstruction: Le hard French, ou, l'architecture française des trente glorieuses*. Paris: Picard, 1988.

"Une vieillesse sacrifiée." *France-Soir*, 1 March 1962.

Villermé, Louis-René. "De la mortalité dans les divers quartiers de la ville de Paris." *Annales d'hygiène publique et de médecine légale* 3 (1830): 294–341.

Vital-Durand, Brigitte. "'Ils en ont profité s'en aller.'" *Libération*, 9 September 2003.

Voldman, Danièle, ed. *Les origines des villes nouvelles de la région parisienne: 1919–1969*. Paris: Institut d'Histoire du Temps Présent, 1990.

Wacquant, Loïc. *Urban Outcasts: A Comparative Sociology of Advanced Marginality*. Cambridge: Polity Press, 2008.

Waintrop, Michel. "Au 'carré des indigents,' des hommes et des numéros." *La Croix*, 3 November 2003.

Walkowitz, Judith. *City of Dreadful Delight: Narratives of Sexual Danger in Late Victorian London*. Chicago: University of Chicago Press, 1992.

Waters, Thomas. "Heat Illness: Tips for Recognition and Treatment." *Cleveland Clinic Journal of Medicine* 68, no. 8 (2001): 685–87.

Weber, Eugen. *The Hollow Years: France in the 1930s.* New York: Norton, 1994.

Weil, Patrick. "Georges Mauco, expert en immigration: Ethnoracisme pratique et antisémitisme fielleux." In *L'antisémitisme de plume 1940–1944: Études et documents,* edited by Pierre-André Taguieff, 267–76. Paris: Berg, 1999.

Weizman, Eyal. "The Politics of Verticality." http://www.opendemocracy.net/conflict-politicsverticality/article_801.jsp.

Wisner, Ben, Piers Blaikie, Terry Cannon, and Ian Davis. *At Risk: Natural Hazards, People's Vulnerability and Disasters.* New York: Routledge, 2004.

Young, Iris Marion. *Inclusion and Democracy.* New York: Oxford University Press, 2000.

Index